Interactive
Mathematics Program®

INTEGRATED HIGH SCHOOL MATHEMATICS

FIRST EDITION AUTHORS:
Dan Fendel, Diane Resek, Lynne Alper, and Sherry Fraser

CONTRIBUTORS TO THE SECOND EDITION:
Sherry Fraser, IMP for the 21st Century
Jean Klanica, IMP for the 21st Century
Brian Lawler, California State University San Marcos
Eric Robinson, Ithaca College, NY
Lew Romagnano, Metropolitan State College of Denver, CO
Rick Marks, Sonoma State University, CA
Dan Brutlag, Meaningful Mathematics
Alan Olds, Colorado Writing Project
Mike Bryant, Santa Maria High School, CA
Jeri P. Philbrick, Oxnard High School, CA
Lori Green, Lincoln High School, CA
Matt Bremer, Berkeley High School, CA
Margaret DeArmond, Kern High School District, CA

Key Curriculum Press

Year 2
Second Edition   I M P

This material is based upon work supported by the National Science Foundation under award numbers ESI-9255262, ESI-0137805, and ESI-0627821. Any opinions, findings, and conclusions or recommendations expressed in this publication are those of the authors and do not necessarily reflect the views of the National Science Foundation.

Key Curriculum Press
1150 65th Street
Emeryville, California 94608
email: editorial@keypress.com
www.keypress.com
10 9 8 7 6 5 4 3 2 1   13 12 11 10 09
ISBN 978-1-60440-030-4
Printed in the United States
of America

Project Editor
Sharon Taylor

Consulting Editor
Mali Apple

Project Administrator
Juliana Tringali

Professional Reviewer
Rick Marks, Sonoma State University, CA

First Edition Teacher Reviewers
Daniel R. Bennett, Moloka'i High School, HI
Larry Biggers, Highlands High School, TX
Dave Calhoun, Fresno Unified School District, CA
Dwight Fuller, Ponderosa High School, CA
Daniel S. Johnson, Silver Creek High School, CA
Brent McClain, Vernonia School District, OR
Amy C. Roszak, Cottage Grove High School, OR
Carmen C. Rubino, Silver Creek High School, CA
Jean Stilwell, Patrick Henry High School, MN
Wendy Tokumine, Farrington High School, HI

First Edition Multicultural Reviewers
Mary Barnes, M.Sc., University of Melbourne, Australia
Edward D. Castillo, Ph.D., Sonoma State University, CA
Joyla Gregory, B.A., College Preparatory School, Oakland, CA
Genevieve Lau, Ph.D., Skyline College, CA
Beatrice Lumpkin, M.S., Malcolm X College, IL
Arthur Ramirez, Ph.D., Sonoma State University, CA

Copyeditor
LeeAnn Pickrell

Interior Designer
Marilyn Perry

Production Editors
Angela Chen, Andrew Jones

Production Director
Christine Osborne

Production Coordinator
Ann Rothenbuhler

Editorial Production Supervisor
Kristin Ferraioli

Compositor
Lapiz Digital Services

Art Editor
Maya Melenchuk

Technical Artists
Kristin Garneau, Natalie Hill, Matthew Perry, Greg Reeves, Lapiz Digital Services

Photo Researcher
Laura Murray

Illustrators
Juan Alvarez, Tom Fowler, Evangelia Philippidis, Sara Swan, Diane Varner, Martha Weston, April Goodman Willy

Cover Designer
Jeff Williams

Printer
RR Donnelley Press, Inc.

Mathematics Product Manager
Tim Pope

Executive Editor
Josephine Noah

Publisher
Steven Rasmussen

FOREWORD

Is There Really a Difference? asks the title of one Year 2 unit of the Interactive Mathematics Program (IMP). "You bet there is!" As Superintendent of Schools, I have found that IMP students in our district have more fun, are well prepared for our State Testing Program in the tenth grade, and are a more representative mix of the different groups in our geographical area than students in other pre-college math classes. Over the last few years, IMP has become an important example of curriculum reform in both our mathematics and science programs.

When we decided in 1992 to pilot the Interactive Mathematics Program, we were excited about its modern approach to restructuring the traditional high school math sequence of courses and topics and its applied use of significant technology. We hoped that IMP would not only revitalize the pre-college math program, but also extend it to virtually all ninth-grade students. At the same time, we had a few concerns about whether IMP students would acquire all of the traditional course skills in algebra, geometry, and trigonometry.

Within the first year, the program proved successful and we were exceptionally pleased with the students' positive reaction and performance, the feedback from parents, and the enthusiasm of teachers. Our first group of IMP students, who graduated in June, 1996, scored as well on PSATs, SATs, and State tests as a comparable group of students in the traditional program did, and subsequent IMP groups are doing the same. In addition, the IMP students have become our most enthusiastic and effective IMP promoters when visiting middle school classes to describe math course options to incoming ninth graders. One student commented, "IMP is the most fun math class I've ever had." Another said, "IMP makes you work hard, but you don't even notice it."

In our first pilot year, we found that the IMP course reached a broader range of students than the traditional Algebra 1 course did. It worked wonderfully not only for honors students, but for other students who would not have begun algebra study until tenth grade or later. The most successful students were those who became intrigued with exciting applications, enjoyed working in a group, and were willing to tackle the hard work of thinking seriously about math on a daily basis.

IMP Year 2 places the graphing calculator in central positions early in the math curriculum. Students thrive on the regular group collaboration and grow in self-confidence and skill as they present their ideas to a large group. Most importantly, not only do students learn the symbolic and graphing applications of elementary algebra, the statistics of *Is There Really a Difference?,* and the geometry of *Do Bees Build It Best?,* but the concepts have meaning to them.

I wish you well as you continue your IMP path for a second year. I am confident that students and teachers using Year 2 will enjoy mathematics more than ever as they experiment, investigate, and discover solutions to the problems and activities presented this year.

Reginald Mayo

Superintendent

New Haven Public Schools

New Haven, Connecticut

NOTE TO STUDENTS

This textbook guides the second year of a four-year program of mathematics learning and investigation. As in the first year, the program is organized around interesting, complex problems, and the concepts you learn grow out of what you'll need to solve those problems.

If you studied IMP Year 1

If you studied IMP Year 1, then you know the excitement of problem-based mathematical study, such as devising strategies for a complex dice game, learning the history of the Overland Trail, and experimenting with pendulums. The Year 2 program extends and expands the challenges that you worked with in Year 1. For instance:

- In Year 1, you began developing a foundation for working with variables. In Year 2, you will build on this foundation in units that demonstrate the power of algebra to solve problems.

- In Year 1, you used principles of statistics to help predict the period of a 30-foot pendulum. In Year 2, you will learn another statistical method, one that will help you to understand statistical comparisons of populations. One important part of your work will be to prepare, conduct, and analyze your own survey.

You'll also use ideas from geometry to understand why the design of bees' honeycombs is so efficient, and you'll use graphs to help a bakery decide how many plain and iced cookies they should make to maximize their profits. Year 2 closes with a literary adventure—you'll use Lewis Carroll's classic *Alice's Adventures in Wonderland* to explore and extend the meaning of exponents.

If this is your first experience with the Interactive Mathematics Program (IMP), you can rely on your classmates and your teacher to fill in what you've missed. Meanwhile, here are some things you should know about the program, how it was developed, and how it is organized.

The Interactive Mathematics Program is the product of a collaboration of teachers and mathematicians who have been working together since 1989 to reform the way high school mathematics is taught. Over 500,000 students and 2,000 teachers have used these materials since the first edition was published in 1996. This second edition reflects the experiences, reactions, and ideas of students and teachers who have used the first edition for many years.

Our goal is to give you the mathematics you need in order to succeed in this changing world. We will present mathematics to you in a manner that reflects how mathematics is used and that reflects the different ways people work and learn together. Through this perspective on mathematics, you will be prepared both for continued study of mathematics in college and for the world of work.

This book contains the tasks that will be your work during Year 2 of the program. As you will see, these problems require ideas from many branches of mathematics, including algebra, geometry, probability, graphing, statistics, and trigonometry. Rather than present each of these areas separately, we have integrated them and presented them in meaningful contexts, so you will see how they relate to each other and to our world.

Although the IMP program is not organized into courses called "Algebra," "Geometry," and so on, you will be learning all the essential mathematical concepts that are part of those traditional courses. You will also be learning concepts from branches of mathematics—especially statistics and probability—that are not part of a traditional high school program.

Each unit in this four-year program has a central problem or theme and focuses on several major mathematical ideas. Supplemental problems at the end of the main material for each unit provide additional opportunities for you to strengthen your understanding of the core material or to explore new ideas related to the unit.

To accomplish your goals, you will have to be an active learner. You will experiment, investigate, ask questions, make and test conjectures, reflect on your work, and then communicate your ideas and conclusions both orally and in writing. You will do some of your work in collaboration with fellow students, just as users of mathematics in the real world often work in teams. At other times, you will work on your own. You will talk about what you are doing and why, and you will present your results to the class.

We hope you will enjoy the challenges that the Interactive Mathematics Program presents. And we hope that your experiences give you a deeper appreciation of the meaning and importance of mathematics.

Dan Fendel   Diane Resek   Lynne Alper   Sherry Fraser

CONTENTS

Do Bees Build It Best?—Area, Volume, and the Pythagorean Theorem

Cookies—Systems of Equations and Linear Programming

Is There Really a Difference?—The Chi-Square Test and the Null Hypothesis

Fireworks—Quadratic Functions, Graphs, and Equations

All About Alice—Exponents, Scientific Notation, and Logarithms

Do Bees Build It Best?

Area, Volume, and the Pythagorean Theorem

Do Bees Build It Best?—Area, Volume, and the Pythagorean Theorem

Bees and Containers

Is the honeycomb an efficient shape for storing honey?

This is the central question you will examine in this unit. You will begin by making and evaluating some other containers as you work to define what is meant by an "efficient" container. You'll also start an investigation about bees themselves.

Thao Nguyen and Joshua Wilson use cubes to measure volume.

Building the Biggest

The Biggest Box

Your main task in this activity is to build the biggest box you can from a single sheet of construction paper. In this activity, *biggest* means "holds the most." *Box* means "a container with four rectangular sides, a rectangular bottom, and no top."

You can cut your construction paper and tape the pieces together in any way you like as long as your final product is a box. If your first attempt doesn't satisfy you, try again. Keep working until you think you have built the biggest box possible.

Beyond Boxes

Once you have built the biggest box you think is possible, try to build a bigger container of a different shape—still from a single sheet of construction paper. This time the shape does not need to have rectangular sides.

The Secret Lives of Bees

The central problem of this unit concerns the honeycombs that bees build for storing honey. In this activity you will learn more about the life of a bee so you can better understand the central problem.

Your task is to write a report on a topic related to bees. You may find it helpful to include drawings in your report. Feel free to express your own opinions and conjectures about why things are the way they are. Be sure, however, to distinguish between opinion and fact. Also list all the sources you use.

What to Put It In?

All sorts of containers are used for everyday goods. Containers are made of different materials, have different sizes and shapes, and are used for different purposes. Look around your home for different kinds of containers. If possible, go to a supermarket or other store for more ideas.

1. What shapes did you find? Sketch the various containers you saw.

2. What kinds of materials were used?

3. What units of measurement were used to indicate how much a container holds?

4. What criteria do you think manufacturers use to decide what kind of container to create for their products?

5. Choose one container and write about one of these topics.
   - How the container could have been designed better
   - Why the manufacturer designed the container the way it did

Area, Geoboards, and Trigonometry

Measurement is an important theme in this unit.

Once you refine the central question of the unit, you will examine how to measure area. Area is one of the fundamental measurement concepts in mathematics.

In developing your understanding of area, you will move gradually from the concrete examples of figures on a geoboard to the more abstract world of triangles and other polygons. As you'll see, right triangles and trigonometry can play a useful role in finding areas.

Jenna Balch and Devin Ledesma use a geoboard to investigate the areas of triangles.

Nailing Down Area

In this activity, the unit of area is the smallest square on the geoboard, such as the one shown at the right.

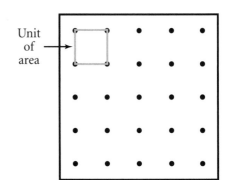

Unit
of
area

1. Construct each figure A to L on a geoboard. Find the area of each figure and record your results.

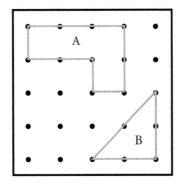

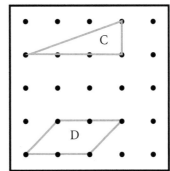

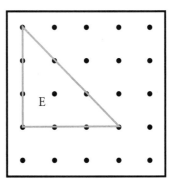

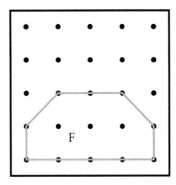

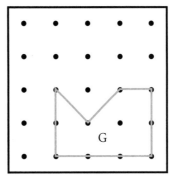

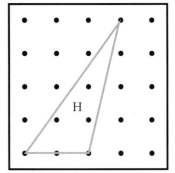

continued ⬛

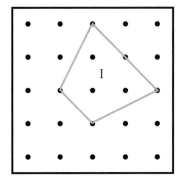 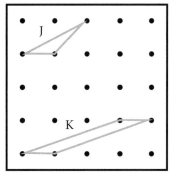 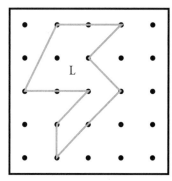

2. Create triangles on the geoboard with an area of half a unit. Find as many different-shaped triangles with this area as you can. Draw the triangles on geoboard paper.

3. Create quadrilaterals (four-sided figures) on the geoboard with an area of exactly 1 unit. Find as many different-shaped quadrilaterals with this area as you can. Record your results on geoboard paper.

Adapted from *About Teaching Mathematics, A K-8 Resource, Third Edition* by Marilyn Burns. Copyright © 2006 Math Solutions Publications. Used by permission.

Approximating Area

For this activity, you will need to trace this figure onto a sheet of paper and then cut out the shape. You may want to make several copies of this figure.

1. Using your copied figure as the unit of area, determine the area of each of these shapes. You will not be able to find the area for the figure in part b exactly, but do the best you can.

a. 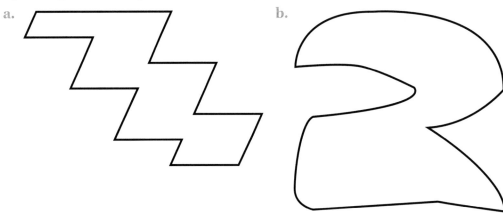 b.

2. Draw a new shape with approximately the same area as the shape in Question 1b.

3. Suppose you have many pieces of cloth that are the shape and size of the unit you used in Question 1. Estimate how many pieces you would need to make a medium-size T-shirt. Explain how you found your estimate.

How Many Can You Find?

As in *Nailing Down Area*, the unit of area in this activity is a basic square on the geoboard, such as the one shown here.

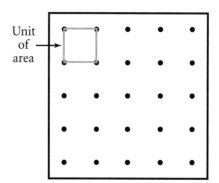

Unit of area →

1. Find the area of each figure A to K.

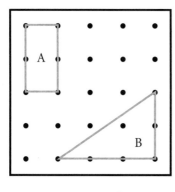

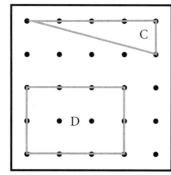

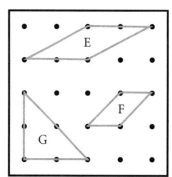

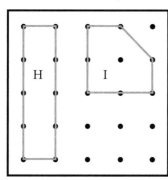

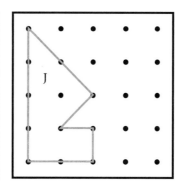

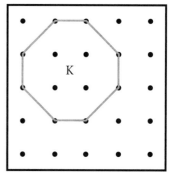

2. On geoboard paper, draw polygons with areas of 6 units. Draw as many different shapes as you can with this area. You'll probably be able to find more shapes if you organize your list in some way.

Adapted from *About Teaching Mathematics, A K-8 Reference, Third Edition* by Marilyn Burns. Copyright © 2006 Math Solutions Publications. Used by permission.

That's All There Is!

Part I: Finding Triangles

In this activity, you will make triangles on the geoboard that fit these conditions.

• Each triangle has an area of 2 units.

• Each triangle has its vertices on pegs.

• Each triangle has a horizontal side.

As in previous activities, the unit of area is the basic square on the geoboard.

Try to find all the possible shapes for triangles that fit these conditions. Keep track of the triangles you find. You will include them in the poster you will make in Part II.

Part II: Studying Triangles

Examine the triangles you found. Sort them into groups according to one or more of the properties they have in common. Then create a poster summarizing your work. Your poster should explain and demonstrate your method of organizing the triangles.

Part III: Further Questions

If you have time, work on these questions.

1. Prove that you found all possible shapes for the triangles in Part I.

2. What areas other than 2 units are possible for a triangle that has its vertices on pegs and has one horizontal side?

3. See if you can make other triangles with an area of 2 units that *don't* have a horizontal side but still have vertices on pegs.

4. Explore Question 3 for triangles with areas other than 2 units.

An Area Shortcut?

Your hand is a complicated shape, so finding the area of a tracing of your hand is not an easy task. In this activity, you'll try it one way and then decide whether a proposed shortcut is a good method.

1. Trace your hand on a piece of graph paper so you get a diagram like the one shown here. Find the area of your hand. Use one grid square as the unit.

2. When Justin Short was asked to measure the area of his hand, he proposed this shortcut.

> I laid a piece of string around the outline of my hand. Then I cut the string to the same length as the outline.

> I reshaped the string into a more convenient rectangle shape. Then I counted the squares inside the rectangle to get the area of my hand.

Try Justin's proposed shortcut. Decide whether you think his method is a good one or not. Explain your conclusion.

Halving Your Way

Look at the rectangle on the geoboard at right.

Your task is to cut this rectangle into two parts with equal area using a rubber band on the geoboard. You may attach the rubber band only at pegs that are on or inside the rectangle. Your goal is to find all possible ways to cut this rectangle into two parts of equal area. (There are exactly 19 ways.) Record your work on geoboard paper.

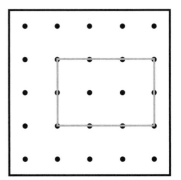

The diagrams here show three methods. The dashed lines represent the rubber band. Although the first two methods are similar, you would count them separately.

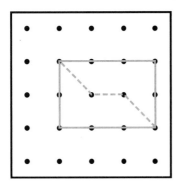 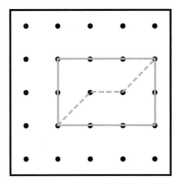 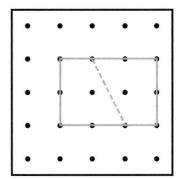

The following method is not allowed because it cuts the rectangle into three parts.

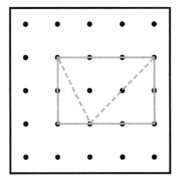

The Ins and Outs of Area

As you may have realized by now, the area of a triangle is related to its **base** and height. Your task now is to find and then explain a rule or formula to describe this relationship.

1. These diagrams show various triangles on geoboards.

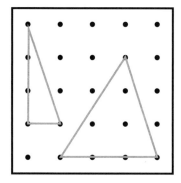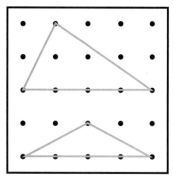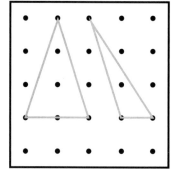

a. Use an In-Out table like the one here to organize information about these triangles. Each row of your table will represent one triangle.

In		Out
Base of triangle	Height of triangle	Area of triangle

b. Create more of your own triangles. Add information about the triangles to your table.

c. Find a rule or formula for your table that expresses the area as a function of base and height.

2. Using a diagram rather than the pattern from the In-Out table, explain why your formula makes sense.

A **parallelogram** is a quadrilateral (a four-sided polygon) in which both pairs of opposite sides are parallel. For example, in quadrilateral *ABCD*, $\overline{AB}$ is parallel to $\overline{CD}$ and $\overline{BC}$ is parallel to $\overline{AD}$, so *ABCD* is a parallelogram.

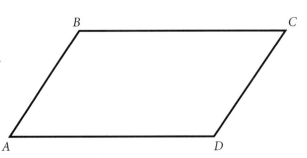

A rectangle is a special kind of parallelogram in which the angles are all right angles. For example, in rectangle *KLMN*, $\overline{KL}$ is parallel to $\overline{NM}$ and $\overline{LM}$ is parallel to $\overline{KN}$, so *KLMN* is also a parallelogram.

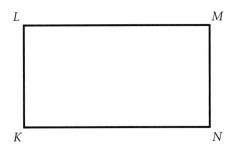

A **trapezoid** is a quadrilateral in which only one pair of opposite sides is parallel. For example, in quadrilateral *RSTU*, $\overline{ST}$ is parallel to $\overline{RU}$ but $\overline{RS}$ is not parallel to $\overline{UT}$, so *RSTU* is a trapezoid.

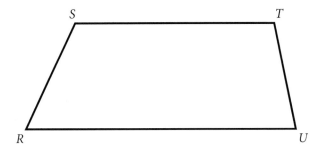

Polygons are often named by listing their vertices sequentially. For polygons with more than three vertices, the sequence in which the vertices are named should reflect how the vertices are connected. For instance, trapezoid *RSTU* has other names, such as *TURS* and *SRUT*. However, this trapezoid cannot be called *STRU* because *T* and *R* are not connected by a side of the figure.

continued ▶

Here is a summary of other geometric notation.

$\overleftrightarrow{AB}$ The *line* through two points, A and B

$\overline{AB}$ The *line segment* from point A to point B

AB The *length* of the line segment from point A to point B

$$\begin{array}{c} \underset{A}{\bullet} \overset{\text{3 centimeters}}{\rule{3cm}{0.4pt}} \underset{B}{\bullet} \end{array} \quad AB = 3 \text{ centimeters}$$

$\overrightarrow{AB}$ The *ray* from point A through point B

$$\underset{A}{\bullet} \xrightarrow{\hspace{2cm} \underset{B}{\bullet} \hspace{2cm}}$$

Going into the Gallery

The Situation

Yoshi is an emerging young artist who is planning a traveling exhibition of his artwork. Because his paintings are quite large, he's concerned about whether they will fit through the doorways of the galleries where they will be displayed.

Each painting is done on a triangle-shaped canvas. When they are delivered to galleries, the paintings must be kept upright to avoid damage. One side of the triangle is used as the horizontal base on which the painting will slide through a doorway.

Your Task

These triangles show the shapes of five of Yoshi's paintings. The triangles use a scale in which 1 centimeter represents 1 meter in the painting. For each triangle, find the height of the lowest possible doorway through which a painting of that shape and size will slide. Also state which side (I, II, or III) should be horizontal.

continued ▶

You will be given a copy of these triangles. You can cut them out and then hold them upright to model sliding them through a doorway.

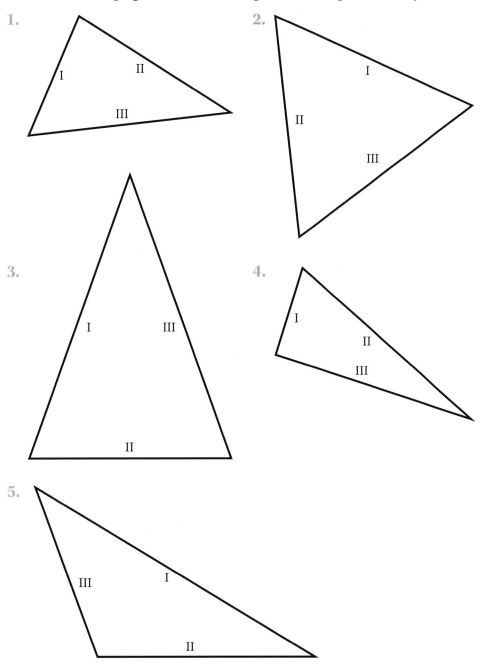

1.

2.

3.

4.

5.

Forming Formulas

You have seen that the area of any triangle is half the product of its base and the **altitude** to that base. For example, the area of this triangle is $\frac{1}{2}bh$.

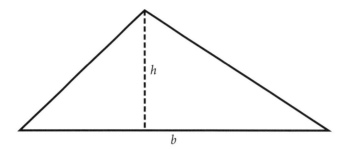

In this activity, your task is to find two formulas—one for the area of a parallelogram and one for the area of a trapezoid. Part of this task is deciding what measurements are important in finding the area of each figure.

Here are two possible approaches.

• Use geoboard paper. Draw examples, find their areas, make a table of data, and look for patterns.

• Make parallelograms and trapezoids out of paper. Look for ways you can cut the figures and fit the pieces together again to make more familiar shapes.

Your write-up should include your formulas, some examples for each formula, and explanations of why the formulas work.

A Right-Triangle Painting

The shape of a certain right triangle particularly appeals to Yoshi's artistic eye. He has done many paintings on triangular canvases, but his favorites all use right triangles with an acute angle of 55°, like the triangle shown here. He thinks that it might be the ratios of the lengths of the sides that appeal to him. He'd like your help in investigating this idea.

1. a. Draw a right triangle *ABC* with a right angle at **vertex** *C* and a 55° angle at vertex *A*, like the one shown. (*Suggestion:* Start by drawing a 55° angle. Extend both sides of the angle until it is at least 10 centimeters long. Then draw a right angle from the end of one side. Extend your lines to complete the triangle.)

   b. Carefully measure and record the lengths of all three sides. Give your measurements to the nearest millimeter.

2. Find each ratio. Remember that *BC* means the *length* of the line segment connecting points *B* and *C*.

   a. $\dfrac{BC}{AB}$ b. $\dfrac{AC}{AB}$ c. $\dfrac{BC}{AC}$

3. Do you think your classmates will get the same results for Questions 1 and 2 that you got? Explain in detail why or why not.

A Trigonometric Summary

The concept of similar triangles allows us to define the trigonometric functions as ratios of the sides of right triangles. This statement summarizes the reasoning.

> If an acute angle of one right triangle is the same size as an acute angle of another right triangle, then the triangles are similar and the ratios of corresponding sides are equal.

Suppose you are given an acute angle. In other words, you are given an angle between 0° and 90°, not including 0° or 90°. To define the trigonometric functions for that angle, you can view it as part of a right triangle.

In $\triangle ABC$, $\angle A$ represents the acute angle you are starting with. $\overline{AC}$ is called the leg *adjacent* to $\angle A$. $\overline{BC}$ is called the leg *opposite* $\angle A$.

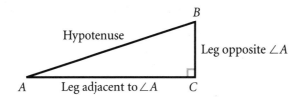

(*Reminder:* $\overline{AC}$ is the adjacent leg when you are thinking about $\angle A$, but it would be the opposite leg if you were thinking about $\angle B$. Also recall that in any right triangle, the longest side is called the **hypotenuse** and the two shorter sides are called the **legs.**)

The **tangent** of $\angle A$ is the ratio of the length of the leg opposite $\angle A$ to the length of the leg adjacent to $\angle A$. The tangent of $\angle A$ is abbreviated tan A. The definition is often written as

$$\tan A = \frac{\text{opposite}}{\text{adjacent}}$$

In $\triangle ABC$

$$\tan A = \frac{BC}{AC}$$

continued ▶

The **sine** of $\angle A$ is the ratio of the length of the leg opposite $\angle A$ to the length of the hypotenuse. The sine of $\angle A$ is abbreviated sin A. The definition is often written as

$$\sin A = \frac{\text{opposite}}{\text{hypotenuse}}$$

In $\triangle ABC$

$$\sin A = \frac{BC}{AB}$$

The **cosine** of $\angle A$ is the ratio of the length of the leg adjacent to $\angle A$ to the length of the hypotenuse. The cosine of $\angle A$ is abbreviated cos A. The definition is often written as

$$\cos A = \frac{\text{adjacent}}{\text{hypotenuse}}$$

In $\triangle ABC$

$$\cos A = \frac{AC}{AB}$$

There are three other ratios of side lengths within a right triangle. These ratios, **cotangent, secant,** and **cosecant,** are used less often than the others. They usually do not have their own calculator keys.

Each of these ratios is the reciprocal of one of the other three ratios. They can be defined as

$$\text{cotangent } A = \frac{1}{\tan A}$$

$$\text{secant } A = \frac{1}{\cos A}$$

$$\text{cosecant } A = \frac{1}{\sin A}$$

They are abbreviated, respectively, cot A, sec A, and csc A.

Just Count the Pegs

Justin Short has a new shortcut. He has a formula to find the area of any polygon on the geoboard that has no pegs in its interior. His formula is like a rule for an In-Out table. The *In* is the number of pegs on the boundary of the polygon. The *Out* is the area of the figure.

Sarah Shorter says she has a shortcut for any geoboard polygon with exactly four pegs on the polygon's boundary. All you have to tell her is how many pegs it has in the interior. She will use her formula to find the area immediately.

Flashy Shortest says she has the best formula yet. If you make *any* polygon on the geoboard and tell her both the number of pegs in the interior and on the boundary of the polygon, her formula will give you the area in a flash!

Your goal is to find Flashy's "superformula." You might begin with her friends' more specialized formulas. Here are some suggestions for how to proceed.

1. Begin by trying to find Justin's formula and some variations, as described in parts a to d.

   a. Find a formula for the area of a polygon with no pegs in the interior. Your formula should use the number of pegs on the boundary as the *In* and give you the area as the *Out*. Make specific examples on the geoboard to get data for your table.

continued

b. Find a different formula that works for polygons with exactly one peg in the interior. Again, use the number of pegs on the boundary as the *In* and the area as the *Out*.

c. Pick a number greater than 1. Find a formula for the area of polygons with that number of pegs in the interior.

d. Do more cases like those in part c.

2. Find Sarah's formula and others like it.

a. First find a formula for the area of polygons with exactly four pegs on the boundary. Your formula should use the number of pegs in the interior as the *In* and give you the area as the *Out*.

b. Pick a number other than 4. Find a formula for the area of polygons with that number of pegs on the boundary. Again, use the number of pegs in the interior as the *In* and the area as the *Out*.

c. Do more cases like those in part b.

When you have finished Questions 1 and 2, look for a superformula that works for all figures. Your formula should have two inputs: the number of pegs in the interior and the number of pegs on the boundary. The output should be the area of the figure.

Try to be as flashy as Flashy!

○ *Write-up*

1. *Problem Statement*

2. *Process:* Explain the methods you used to find your formulas.

3. *Solution:* Give all the formulas you found, along with examples showing that they work.

4. *Self-assessment*

Each Problem of the Week (POW) is unique, so the form of the write-up will vary from one POW to the next. Nevertheless, most of the categories you will use for your POW write-ups will be the same throughout the year. Here is a summary of the standard categories.

Some POW write-ups will use other categories or require more specific information within a particular category. But if the write-up instructions for a given POW simply list a category by name, use these descriptions.

1. *Problem Statement:* State the problem clearly and in your own words. Your problem statement should be clear enough that someone unfamiliar with the problem could understand what you are being asked to do.

2. *Process:* Describe what you did in attempting to solve the problem. Use your notes as a reminder. Include things that didn't work out or that seemed like a waste of time. Do this part of the write-up even if you didn't solve the problem.

   If you get assistance of any kind on the problem, tell what the assistance was and how it helped you.

continued ▶

3. *Solution:* State your solution as clearly as you can. Explain how you know that your solution is correct and complete. If you obtained only a partial solution, give that. If you were able to generalize the problem, include your general results.

   Write your explanation in a way that will be convincing to someone else—even someone who initially disagrees with your answer.

4. *Extensions:* Invent some extensions or variations to the problem. That is, create some related problems. They can be easier, harder, or at a similar level of difficulty as the original problem. You don't have to solve these additional problems.

5. *Self-assessment:* Tell what you learned from this problem. Be as specific as you can. Also assign yourself a grade for your work on this POW, and explain why you think you deserve that grade.

More Gallery Measurements

To help Yoshi plan for his tour, several friends measured his paintings. They drew triangles and marked the measurements on each diagram. Unfortunately, they weren't thinking about the doorways and the altitudes, so their measurements aren't necessarily the most useful.

The diagrams below show his friends' measurements. One of the altitudes in each triangle is shown as a dashed line labeled *h*.

Your task is to use one of the trigonometric functions—sine, cosine, or tangent—to find the value of *h* in each case. Base your work on the measurements shown, not on any measurements you make yourself.

1.

94 inches

h

30°

2.

h

43°

2.7 meters

3.

61°

430 centimeters

h

4.

h

58°

14 feet

Sailboats and Shadows

1. A sailboat is in trouble. The people on board are considering trying to swim to shore.

   A lookout station on shore tells the people that they are 2.3 miles from the station and that the line from the station to the boat forms an angle of 35° with the shoreline. (Assume the shoreline is straight, as shown in the diagram.)

   The people are capable of swimming 1.5 miles. Will they be able to make it to shore or should they call for help? Explain.

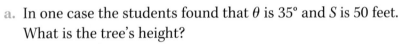

2. Some students were using shadows to find the heights of trees near their school. They used the diagram shown here to represent the situation. In this diagram, θ (the Greek letter "theta") represents the angle of elevation of the sun. S represents the length of the tree's shadow.

   a. In one case the students found that θ is 35° and S is 50 feet. What is the tree's height?

   b. Later that day, with a different tree, they measured θ to be 60° and S to be 20 feet. What is the height of that tree?

   c. Develop a general expression for the height of a tree in terms of S and θ.

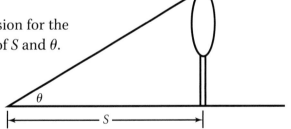

A Special Property of Right Triangles

Right triangles are all around us. The special properties of right triangles have interested people for many centuries.

In the next several activities, you will discover, prove, and apply a property of right triangles that is one of the most famous principles in all of mathematics. This principle was known to civilizations throughout the ancient world. Today it is usually associated with the name of a Greek mathematician.

Lori Green helps Alice Lenz, Robert Corral, Christopher Refuerzo, and Kelsey Coria discover a special property of triangles.

Tri-Square Rug Games

A rug designer decided to make a rug consisting of three separate square pieces sewn together at their corners, with an empty triangular space between them.

The rug was an immediate hit, so the designer decided to make more of them. He called these creations *tri-square rugs*. One tri-square rug is shown here.

Al and Betty thought these tri-square rugs could be used to make a great game. They made up these rules.

Let a dart fall randomly on a tri-square rug.

- If it hits the largest of the three squares, Al wins.

- If it hits either of the other two squares, Betty wins.

- If the dart misses the rug, simply let another dart fall.

Your goal is to decide which tri-square rugs you would prefer if you were Al and which you would prefer if you were Betty. You will also explore whether any rugs lead to a fair game.

1. You will be given three sheets of chart paper. Label the three sheets "Fair Game," "Al Wins in the Long Run," and "Betty Wins in the Long Run."

2. Use the squares provided by your teacher to make some sample tri-square rugs. For each tri-square rug you make, decide what would happen in Al and Betty's game in the long run. Then carefully post the rug on the appropriate chart.

3. When you have several examples for each category, look for a pattern in your results. Your goal is to find a way to tell who will win in the long run simply by glancing at a tri-square rug.

How Big Is It?

1. Summarize what you have learned so far about area. Discuss both what area means and formulas for computing area.

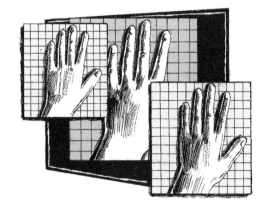

2. Measurement plays an important role in this unit. For example, you measured the lengths of the sides of right triangles, found areas of figures on the geoboard, and compared volumes of boxes built from construction paper.

   What does it mean to measure length, area, and volume? What are the relationships among the units used for each of these measurements? What do these measurements have in common? How are they different?

Any Two Sides Work

The **Pythagorean theorem** expresses an important relationship among the lengths of the sides of a right triangle. This activity gives some examples of the many ways this theorem can be used.

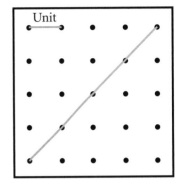

1. This diagram shows the longest diagonal on your geoboard.

    a. Estimate the length of this diagonal. As usual, use the distance between adjacent pegs as the unit of length.

    b. Find the exact length of this diagonal using the Pythagorean theorem.

2. An 8-foot ladder is leaning against a wall, as shown in this diagram. The bottom of the ladder is 2 feet from the wall. How high up the wall does the ladder reach?

3. Marlene wants to check that her door frame makes right angles at the corners. The door is 2.5 meters high and 1.5 meters wide. How long should each diagonal of the door be if the corners are right angles?

4. José thinks the triangle below might be a right triangle. Find the length of each side of this triangle. Use your answers to determine with certainty whether the triangle is a right triangle. Explain your reasoning.

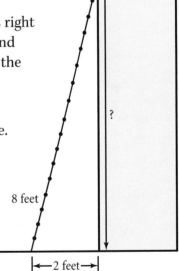

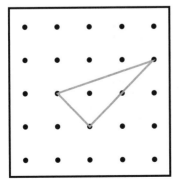

Activity

Impossible Rugs

One morning the rug designer chose three square rugs and found it impossible to make a tri-square rug from them. He could not connect them at the corners and leave a triangular space in the middle. Perhaps this happened to you when you worked on *Tri-Square Rug Games*.

Now the designer wants to know what combinations of squares he *can* use. That is, he wants to know what combinations of squares can be connected at the corners to produce a rug with a triangular space in the middle.

1. List some combinations of square rugs that can be used. Also list some combinations that can't be used.

2. Find a rule that will help the designer determine whether three given squares can form a tri-square rug without actually putting them together. Use your examples as a guide. Explain your answer.

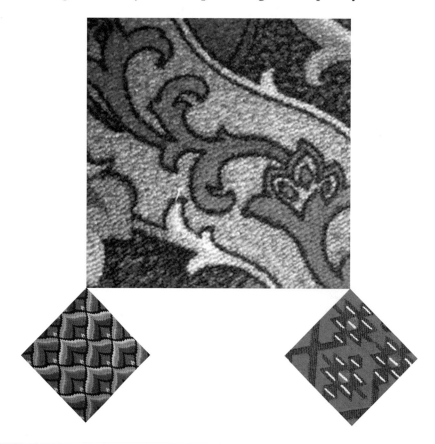

Make the Lines Count

Lots of line segments can be found on a geoboard. The segment labeled *d* in the diagram is one example. In this activity, you will investigate the lengths of such segments.

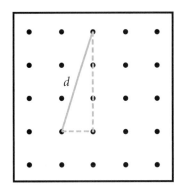

You can find the length of segment *d* by thinking of it as the hypotenuse of a right triangle. You can form such a triangle by drawing the segments shown as dashed lines. As usual, the unit of length is the distance between adjacent pegs.

1. Make diagrams to illustrate all the possible lengths of line segments on the geoboard. Consider only segments that start and end at pegs. Make one diagram for each length you find.

2. Using the Pythagorean theorem, find each length you showed in your answer to Question 1. For any length that is not a whole number, write it in two ways.

   • As a square **root**

   • As a decimal rounded to the nearest hundredth

   For example, you can find the length *d* in the diagram using the equation $1^2 + 3^2 = d^2$. You would then give length *d* as both $\sqrt{10}$ and 3.16.

Proof by Rugs

Al and Betty have another game. They began with this right triangle, which has legs of lengths *a* and *b* and a hypotenuse of length *c*.

Then they made the two square rugs shown here. Each rug has sides of length *a* + *b*. The four triangles in each rug are the same as the single right triangle shown at the right.

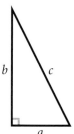

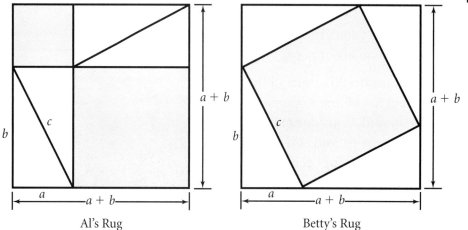

Al's Rug   Betty's Rug

When it is Al's turn, a dart drops on the square rug on the left. If the dart hits the shaded area, Al earns 1 point. When it is Betty's turn, a dart falls on the square rug on the right. If the dart hits the shaded area, Betty earns 1 point. The darts always hit the rugs, and they land randomly. In other words, all points on a rug have the same chance of being hit.

1. Is this a fair game? In other words, is the chance of a dart hitting the shaded area the same for the two rugs? Explain your answer.

2. How do the two rugs demonstrate that the Pythagorean theorem holds true in general?

The Power of Pythagoras

The Pythagorean theorem can be used to find the lengths of many kinds of things.

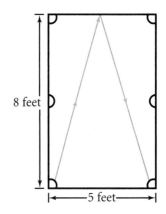

1. Bonny was doing one of her favorite trick billiard shots. As illustrated in the diagram, her shot started at one corner of the table. It hit the exact center of the back cushion and rebounded into the other corner. How far did her billiard ball travel?

2. The scene is a football field. Ben catches a kick and follows the path shown in the diagram. He starts at the east end of one goal line (near the lower right of the diagram). He runs first to the 20-yard line on the west sideline and then to the 50-yard line on the east sideline. Finally he runs for a touchdown at the west end of the other goal line. The field is 53.33 yards wide and 100 yards long. How far does Ben run altogether?

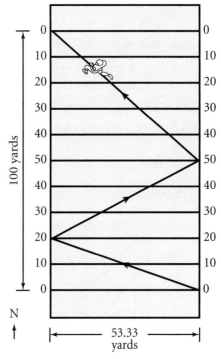

3. Corinne and Deanna decide to race from one corner of an open field to the other. The field is rectangular, 60 meters long and 80 meters wide.

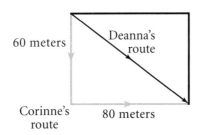

Since Corinne is older and faster, she will run along the outside of the field. Deanna will take the diagonal route. The arrows in the diagram indicate their paths. If Deanna can run at a rate of 5 meters per second, how fast will Corinne have to run to reach the corner at the same time as Deanna?

Tessellation Pictures

People around the world have long been fascinated with shapes that can be used for tessellations. A **tessellation** shape has the property that multiple copies can be fit together to meet these conditions.

- No two copies of the shape overlap.
- There are no gaps between copies.
- You can always fit more copies in any direction.

This POW is about tessellation shapes. It has three parts.

Part I: Make a Tessellation Shape

Step 1: On an index card, draw some kind of path from the upper-left corner to the upper-right corner. Use curves, line segments, or any combination of them. Your path should have no "breaks" or "loops" in it.

It will look something like the drawing at the right.

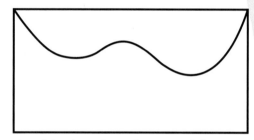

A path has been drawn from upper left to upper right.

Step 2: Cut your index card along the curves or lines you drew. This will create two pieces, as shown below.

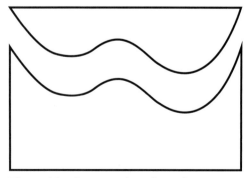

The card has been cut along the path, creating two pieces.

continued ▶

Step 3: Tape the two parts together. The original top edge of the upper part of the card will be along the bottom edge of the other piece.

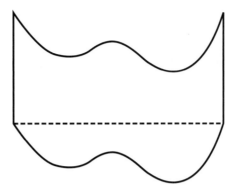

The two parts of the index card have been taped together. This shape is more interesting than a rectangle.

Step 4: Now draw a path from what used to be the upper-left corner of the index card to what used to be the lower-left corner. Cut and tape as shown here.

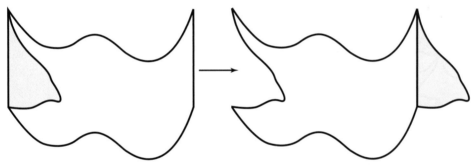

The colored part of the figure on the left has been cut out and taped as shown on the right.

The final tessellation shape looks like this.

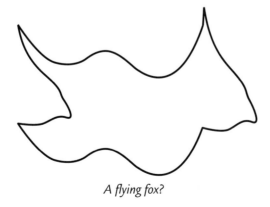

A flying fox?

continued ▶

○ *Part II: Make Some More Tessellation Shapes*

To make a tessellation shape, you don't have to start with a rectangle. You can use the method in Part I with any parallelogram. You can also start with any tessellation shape, but the cutting and taping process will be more complicated.

Make some more tessellation shapes. Don't limit yourself to working with index cards.

○ *Part III: Make a Tessellation Picture*

Choose one of your tessellation shapes. Keep working on it until you create a well-designed, interesting shape to use in your final product.

Now trace many copies of your shape, fitting them together. For example, if your shape looks like the figure labeled "A flying fox?" in Part I, your drawing might look like this.

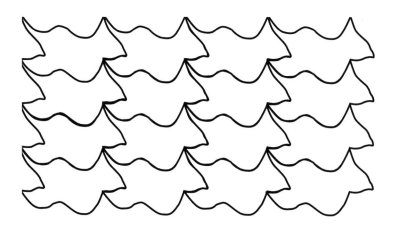

Color each shape to make it interesting. You do not have to color all the copies of your shape the same way.

This product is your tessellation *picture*. It should include at least 10 copies of your tessellation *shape*.

You will hand in this tessellation picture, as well as one copy of each of the other tessellation shapes you make. You will be evaluated primarily on the quality of your tessellation picture, in terms of both its mathematical content and its beauty.

Some IMPish Tessellations shows examples of tessellations that students have made for this POW.

Some IMPish Tessellations

Students in IMP classes at Santa Cruz High School in Santa Cruz, California, and Tamalpais High School in Mill Valley, California, created these tessellation pictures.

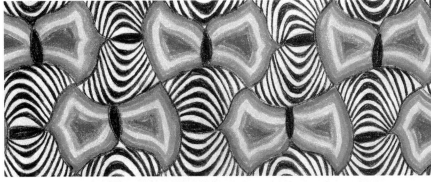

Leslie's Fertile Flowers

Leslie, the landscape architect, has designed a flower bed for an important client. The flower bed will be in the shape of a triangle. It will have sides of lengths 13 feet, 14 feet, and 15 feet, as shown in the diagram.

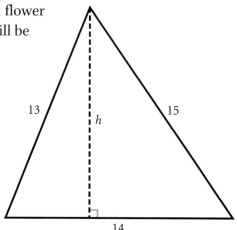

To order the correct amount of fertilizer, Leslie needs to know the area of the flower bed.

Suppose you are Leslie's assistant. To find the area, you need to find the length of the altitude (labeled h).

1. Find the length of the altitude. Guess ways that the 14-foot side might be split into two parts, and try to figure out h from that.

2. Explain how you know that your answer to Question 1 is correct. In other words, how are you sure you made the correct guess?

3. Calculate the area of the flower bed for Leslie.

Flowers from Different Sides

When you gave Leslie the area of the flower bed, she asked, "What if you looked at the triangle from a different point of view? Would the area be the same?" You want to show Leslie how confident you are about your answer. You draw the triangle again, this time using the 15-foot side as the base. You use k for the length of the new altitude.

1. a. Label the two parts of the new base y and $15 - y$. Using the Pythagorean theorem, write two equations that you can use to find the value of k.

   b. Use your equations to find the value of k.

   c. Find the area of the triangle using this base and altitude.

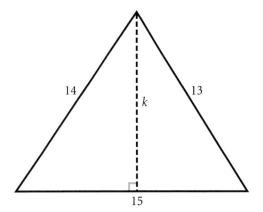

2. Explain to Leslie why you knew that the area would come out the same as before.

3. Find the length of the altitude to the 13-foot side. Check the triangle's area using this base and altitude as well.

The Corral Problem

Rancher Gonzales is trying to decide what shape to use for her corral. Everybody has advice for her.

What do you think? Making a good choice will require an understanding of area, perimeter, altitudes, angles, and **trigonometry.**

Jessica Sanford and Cameron Savage put finishing touches on their tessellations POW.

Don't Fence Me In

Rancher Gonzales is building a corral for her horses. She decides she can afford 300 feet of fencing. For aesthetic reasons, she wants to build the corral in the shape of a rectangle. She also wants as much space as possible inside the corral for her horses to move around.

1. Experiment. Try various lengths and widths for the rectangle. Use your intuition to guess what the corral's dimensions should be.

2. Now make an In-Out table in which the *In* is the width of the rectangle and the *Out* is the area of the rectangle. Make several choices for the width, and find the corresponding areas. Pay attention to how you calculate the areas.

3. Use your work in Question 2 to find a formula or rule for your In-Out table. Use the variable *w* to represent the width of the rectangle and the variable *A* to represent the area.

Rectangles Are Boring!

Just as rancher Gonzales decided to build a square corral, her nephew Juan stopped by for a visit. Juan thinks a square corral is a boring idea. "Why always a rectangle?" he asks. "Why not be innovative? How about a triangle?"

1. Suppose rancher Gonzales uses her 300 feet of fencing to build a corral in the shape of an equilateral triangle. What will the corral's area be?

2. How does that area result compare to building a square corral with 300 feet of fencing?

More Fencing, Bigger Corrals

Rancher Gonzales goes to the supply store to buy the fencing for a square corral. To her surprise, the store is having a half-price sale. She really loves her horses, so she considers spending the amount of money she had originally planned to spend. In other words, she might buy 600 feet of fencing instead of 300 feet. This would allow her to build a bigger corral.

1. a. What would be the area of the square corral if rancher Gonzales used 600 feet of fencing?

   b. How does your answer compare to the area of a square corral made from 300 feet of fencing? In other words, what does doubling the perimeter do to the area?

2. What would be the area of an equilateral-triangle corral if rancher Gonzales used 600 feet of fencing? How does that compare to the area of an equilateral-triangle corral made from 300 feet of fencing?

3. What would be the areas of the square and triangular corrals if rancher Gonzales used 900 feet of fencing to build each one?

4. What generalizations can you make from your results in Questions 1 to 3? It might help to make an In-Out table.

More Opinions About Corrals

Nephew Juan is not the only person giving rancher Gonzales advice. When friends hear she is building a corral, they all have suggestions. When she tells them about her experience with triangles and rectangles, they all agree she should consider only **regular polygons.**

Her niece Juanita thinks a regular pentagon would be a lovely shape for a corral. She wonders if it might even make a larger corral than a square. Your task is to find the area of a corral built in the shape of a regular pentagon that uses 300 feet of fencing.

Simply Square Roots

Your work with the Pythagorean theorem has led to many answers involving square roots. Now you will investigate some general questions about the square-root function. Then you will apply your results to specific problems.

Investigate Questions 1, 2, and 3 by testing specific numbers and looking for general principles. In other words, choose specific numbers for a and b in each case, and see whether the property is true or not. For example, in Question 1, you might find the values of $\sqrt{7 + 10}$ and $\sqrt{7} + \sqrt{10}$ and compare them. Then try other examples.

For each question, state any general conclusions you reach.

1. Is the square root of a sum equal to the sum of the square roots? In other words, is $\sqrt{a + b}$ the same as $\sqrt{a} + \sqrt{b}$?

2. Is the square root of a product equal to the product of the square roots? In other words, is $\sqrt{a \cdot b}$ the same as $\sqrt{a} \cdot \sqrt{b}$?

3. Is the square root of a quotient equal to the quotient of the square roots? In other words, is $\sqrt{\frac{a}{b}}$ the same as $\frac{\sqrt{a}}{\sqrt{b}}$?

4. Write each square-root expression in a different way. Explain your answer in terms of your results for Questions 1, 2, and 3.

    a. $\sqrt{25 \cdot 3}$ b. $\sqrt{49 \cdot 5}$

    c. $\sqrt{18}$ d. $\sqrt{\frac{9}{4}}$

    e. $\sqrt{\frac{3}{16}}$ f. $\sqrt{\frac{25}{7}}$

Building the Best Fence

Now rancher Gonzales is really perplexed. She thought squares were good, but now she sees that pentagons are better. She's wondering if another shape is even better than pentagons.

Based on what she's discovered about rectangles and triangles—and for simplicity and artistic reasons—she decides to consider only regular polygons for her corral. This still gives her a lot of choices.

1. Choose a value greater than 5 for the number of sides of the corral. What would the area of the corral be if rancher Gonzales built it in the shape of a regular polygon with this many sides? (She has decided to use only 300 feet of fencing.)

2. Repeat Question 1 for another regular polygon with more than five sides.

3. Generalize the process you used in Questions 1 and 2. That is, suppose you have a polygon with n sides and a perimeter of 300 feet. (A polygon with n sides is called an n-gon.) Develop a formula for the area of the corral in terms of n.

Falling Bridges

Thorough Ted is a construction engineer. He correctly computes that the maximum safe load of a bridge being planned would be $1000(99 - 70\sqrt{2})$ tons.

Speedy Sam is the safety supervisor. He is asked to design a sign to tell drivers how much weight the bridge can safely hold. He begins with Thorough Ted's expression. Using 1.4 as an approximation for $\sqrt{2}$, he creates the sign based on his calculations.

The bridge opens to traffic on a bright July morning. One hour later, it collapses under a load less than a tenth of the weight shown on Sam's sign!

Speedy Sam tells the city council that he had simply used Ted's figures. Thorough Ted reports that he has been over and over his figures and can't see how they could be wrong.

Write a clear explanation for the city council of why the bridge collapsed.

Adapted from *Mathematics Teacher* by the National Council of Teachers of Mathematics, Vol. 82, No. 9 (December 1989).

Leslie's Floral Angles

Leslie isn't finished with her flower bed from *Leslie's Fertile Flowers*. But thanks to your help, she knows it should look like this diagram.

Leslie has finally begun to build the flower bed. But now she realizes that she needs to know the angles at the corners of the triangle.

Can you use some trigonometry to calculate these for her? Find all three angles to the nearest degree.

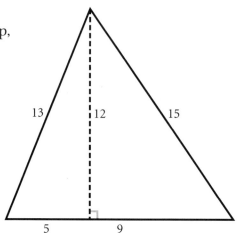

From Two Dimensions to Three

Now you'll move from the two-dimensional world of polygons to the three-dimensional world of solid figures. Area continues to play a role here because solid figures have surfaces too. You'll also learn about volume. You will look at such questions as "What is volume?" "How do you measure volume?" and "How is volume related to surface area?"

Molly Bergland and Libby Hobbs discuss which cylinder in "Which Holds More?" has the greater volume.

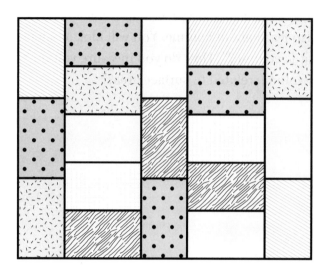

Keisha is making a patchwork quilt. Her quilt will be sewn from rectangular patches of material. Each patch is 3 inches by 5 inches.

Rummaging in the attic, Keisha finds a box of material that her grandmother saved for sewing projects. She pulls out a rectangular piece of satin 17 inches wide and 22 inches long. As you can imagine, Keisha wants to cut as many patches as she can from the satin. Each patch must be a single section of the material. She won't sew scraps together to make patches.

1. How many 3-by-5-inch patches can Keisha get from the 17-by-22-inch piece of satin? Draw a diagram proving your answer.

2. How many 9-by-10-inch patches could she get from this 17-by-22-inch piece of satin? Can you prove your answer? How many 5-by-12-inch patches? How many 10-by-12-inch patches?

3. Suppose Keisha found a piece of satin 4 inches wide and 18 inches long. How many 3-by-5-inch patches could she get from this piece? What if the satin were 8 inches by 9 inches?

continued

Begin with the specific situations described in Questions 1 to 3. Then experiment with other patch sizes and other sizes for the piece of satin. Your write-up should give your answers to the specific questions and describe any other results you found.

○ *Write-up*

1. *Problem Statement*

2. *Process*

3. *Results:* Include diagrams to justify your answers to the questions and to explain any general observations.

4. *Evaluation*

5. *Self-assessment*

Adapted from *Mathematics Teacher* by the National Council of Teachers of Mathematics, Vol. 82, No. 8 (November 1989).

Flat Cubes

A **net** is a flat pattern that can be cut out and folded to make a solid shape.

The diagram here shows a pattern that can be cut out and folded into the shape of a cube. The flat pattern is a net for the cube.

Your task is to sketch at least four different nets that could be cut out and folded into a cube with 1-inch sides. At least one of your nets should *not* have four squares in a straight line. Use dashed lines to show where to fold the nets.

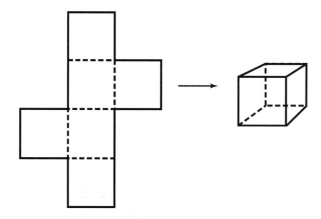

Flat Boxes

Each diagram shows a box, or rectangular solid. The diagrams are not drawn to scale, so use the measurements shown.

1. Make a net that will fold into each of the given solids. Your folded nets should include the top and the bottom of the solids.

2. Cut out and fold your nets to make each rectangular solid.

3. Find the total **surface area** of each rectangular solid.

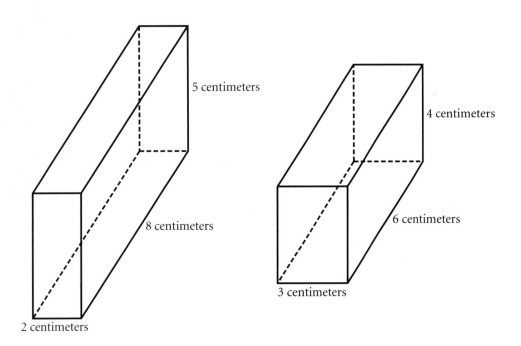

Not a Sound

Pearl was tired of hearing "Turn that music down!" every evening. She decided to soundproof her room so she could listen to her music uninterrupted.

Soundproofing material is sold in flat sheets that can be attached to walls. Pearl decided to put soundproofing on the floor and ceiling too, just to be on the safe side.

1. What if you want to soundproof a room where you live? Choose a box-shaped room in your house. Figure out how many square feet of soundproofing material you would need to soundproof that room. Explain the process you used to find your answer.

2. Explain how this problem relates to the unit problem about bees and their honeycombs.

A Voluminous Task

1. Working with the cubes provided for you, build each of the ten solid figures shown. Assume there are no hidden stacks of cubes. In other words, assume the back edges of each figure drop straight down.

2. a. Find the surface area and the **volume** of each solid figure. Use a single cube as the unit of volume and a face of that cube as the unit of area.

   b. Write a description of how you found the volumes and surface areas. If you used any shortcuts, explain them.

continued ▶

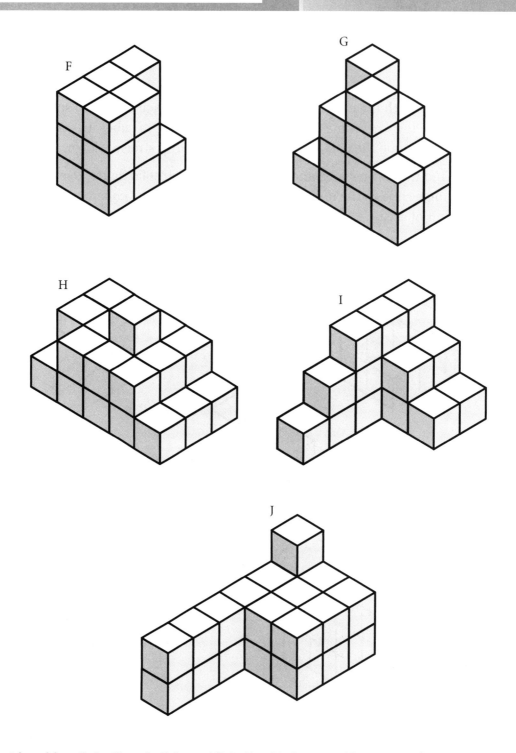

Adapted from *Seeing Shapes* by E. Ranucci (Palo Alto, CA: Creative Publications, 1973).

Put Your Fist into It

Part I: Fists and Volume

When you take measurements, sometimes you have to improvise. For instance, a gardener may "pace off" the dimensions of a lawn, using his or her stride as the unit of length.

Volumes are usually measured with units like cubic centimeters or cubic feet. In this activity, you will use an improvised unit of volume: your fist.

1. Choose three objects. Estimate the volume of each object, using your fist as the unit of volume.

2. Estimate the number of cubic inches in your fist.

3. Use your answer to Question 2 to estimate the number of cubic inches in each of your three objects.

Part II: A Triangular Puzzle

You will need to put various geometric ideas together to solve this puzzle. In the diagram, notice that $\overline{BD}$ is perpendicular to $\overline{AC}$, $\angle BAC$ is 15°, and $\angle BDC$ is 25°. Also note that AE is 75 feet and DE is 40 feet.

Find BC to the nearest foot. Explain your reasoning.

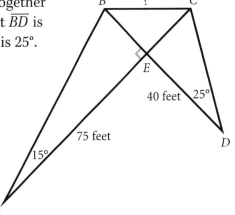

The Ins and Outs of Boxes

In *Flat Boxes,* you found the surface areas for two rectangular solids. To find those surface areas, you needed to know the dimensions of the solids: their lengths, widths, and heights.

Now you will look at how to calculate both the surface area and the volume of a rectangular solid when you know its length, width, and height. Use a centimeter as the unit of length, a square centimeter as the unit of surface area, and a cubic centimeter as the unit of volume.

1. Enter the information from *Flat Boxes* in an In-Out table like this one.

In			Out
Length	Width	Height	Surface area

2. Find the volume of each figure from *Flat Boxes.* Enter this information in an In-Out table like the one here.

In			Out
Length	Width	Height	Volume

3. Find rules for each of your In-Out tables. Use the letters *l*, *w*, and *h* to represent length, width, and height. Use *S* and *V* to represent surface area and volume. Your goal is to express *S* and *V* as functions of *l*, *w*, and *h*. You may want to examine more rectangular solids to add more rows to your tables.

4. Justify any formulas you found in Question 3.

A Sculpture Garden

Emma's neighbor, Richard, bought a sculpture to put in his garden. Emma loved how beautiful the sculpture looked. She was inspired to construct her own outdoor sculpture.

Emma's father gave her eight wooden crates, all cubes of the same size. Emma decided to paint the crate cubes a bright color and stack them.

Emma realized she could save materials and time by painting only the parts of the cubes that were exposed. This would include the parts of the cubes that would touch the ground to keep them from rotting. She wondered how she could stack the eight cubes to require the least amount of paint.

1. Find a way to arrange Emma's eight cubes that would require the least amount of paint. Sketch this arrangement. Can you find another arrangement that is just as economical?

2. Discuss how this situation is related to the unit problem about bees and their honeycombs.

The World of Prisms

You may have seen a glass object called a *prism* break up light into a rainbow-like pattern. A **prism** is also a special type of solid geometric figure. The object used to break up light is a special example of this type of solid.

Geometrically, a prism is formed by moving a polygon through space for a fixed distance, keeping it parallel to its original position. The initial and final positions of the polygon represent the **bases** of the prism. The perpendicular distance between the bases is called the **height** of the prism.

Here are some examples of prisms. The shaded face is one of the bases in each case.

A prism is classified by the shape of its base. Thus example A is called a triangular prism, B is a hexagonal prism, and C is a rectangular prism.

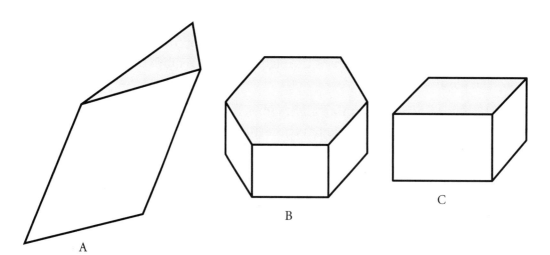

continued ▶

If the direction through which the base has moved is perpendicular to its original position, the resulting solid is called a **right prism.** An ordinary box is an example of a right rectangular prism. Examples B and C are intended to show right prisms. A prism that is not a right prism is called an oblique prism. Example A is intended to represent an oblique prism.

The faces of a prism other than its bases are called **lateral faces.** (The word *lateral* means "side.") Because of the way a prism is defined, all of the lateral faces are parallelograms, no matter what the shape of the base. If the prism is a right prism, the lateral faces are all rectangles. The sum of the areas of the lateral faces is called the **lateral surface area** of the prism.

The line segment connecting a vertex of the lower base to the corresponding vertex of the upper base is called a **lateral edge.**

Shedding Light on Prisms

Part I: Building and Measuring

In this activity, you will build right prisms out of cubes. Questions 1 to 5 describe different prisms, using a picture or other information to describe the prism's base. In each case, the base is a polygon.

For each prism, find the value of each of these measurements.

a. Height of the prism

b. Area of the base

c. Volume of the prism

d. Perimeter of the base

e. Lateral surface area of the prism

Use the edge of a cube as the unit of length. Use the face of a cube as the unit of area. Use the cube itself as the unit of volume.

Your goal in Part II will be to discover some general relationships among the five measurements listed. Think about this goal as you work on the individual prisms.

1. A right prism that has the polygon at the right for its base and is 4 units high

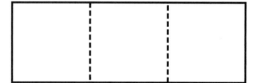

2. A right prism that has the polygon at the right for its base and is 6 units high

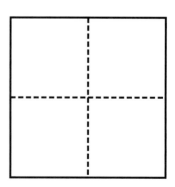

continued ▶

3. A right prism that has the polygon at the right for its base and is 3 units high

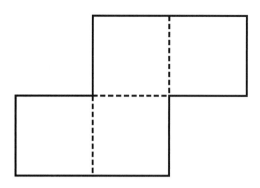

4. A right prism that is 9 units high and has a rectangular base 6 units in area

5. A right prism that is 6 units high and has a nonrectangular base 6 units in area

Part II: Finding Formulas

Study your answers to the questions in Part I, and create more prisms if necessary. Then find general rules or formulas that show relationships among the measurements you found.

Pythagoras and the Box

1. Peter bought a special pen as a birthday present for an artist friend. He has a sturdy box he wants to use to mail the pen. The box is 4 inches wide, 2 inches deep, and 8 inches high.

   The pen is 10 inches long. Peter knows he will have to place it in the box along the "long" diagonal—that is, along the line segment in the diagram connecting point *A* to point *B*.

   Will the pen fit in this box? (Ignore the thickness of the pen.) Give the length of the long diagonal, and explain how you found the answer.

   Begin by finding the length of $\overline{AC}$ using right triangle *ADC*. Then use the fact that $\angle ACB$ is a right angle. You may want to use a shoe box or another box-shaped object to help you picture the situation.

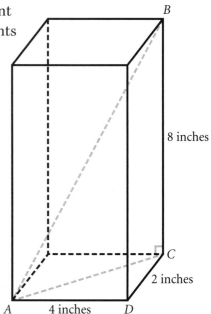

8 inches

2 inches

4 inches

2. a. Find a box-shaped object around the house. Measure its length, width, and height.

   b. Use your measurements to compute the length of the long diagonal of your object.

   c. If possible, confirm your answer to part b by measuring the long diagonal.

3. Generalize your findings from Questions 1 and 2 to a box with dimensions *p*, *q*, and *r* units. In other words, find the length of the segment from point *U* to point *V*, as shown here. Explain your answer.

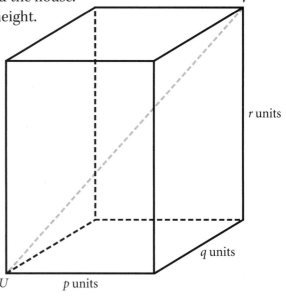

r units

q units

p units

Back on the Farm

1. A Long Drink

Farmer Minh, a neighbor of rancher Gonzales, has a drinking trough for his animals. The trough is in the shape of the triangular prism shown here. The triangle that forms the base of this prism is a right triangle with 1-foot-long legs. The trough is 5 feet long.

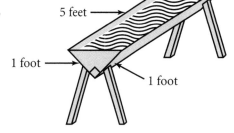

How much water will the trough hold when it is full? Give your answer in cubic feet.

2. Farmer Minh's Barn

Farmer Minh has been listening to his neighbor discuss corral shapes. He is thinking about how he can apply her ideas to the barn he is building. The barn will be in the form of a prism. Its base will be a regular polygon, something like the prism in this diagram.

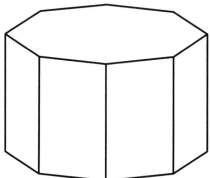

The sides, of course, will be vertical. He wants the base of the prism to have a perimeter of 300 feet (the same as the perimeter of his neighbor's corral).

Farmer Minh is trying to decide whether to make the barn floor (the prism's base) in the shape of a regular octagon (8 sides), a regular decagon (10 sides), or a regular dodecagon (12 sides). He will want to paint the outside of the barn, including the doors. So he decides to look for the shape that will give the barn the least lateral surface area. The barn's walls will be 10 feet tall.

1. Before doing any computations, guess which shape for the barn floor will give the least lateral surface area. Write down your guess.

2. Now do some computations. Find the lateral surface area of the barn for each of the three shapes farmer Minh is considering. Explain your results.

3. How do your results compare to your guess?

Which Holds More?

Take two identical sheets of binder paper ($8\frac{1}{2}$ inches by 11 inches). Make one into a tall, skinny cylinder by taping the two 11-inch sides together where they meet. Make the other into a short, fat cylinder by taping the two $8\frac{1}{2}$-inch sides together where they meet. The base in each case is a circle. You will form two objects like those shown here.

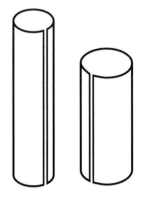

Now place your cylinders on a flat surface like a table. The flat surface will act as the bottom of the cylinders.

1. Imagine filling your cylinders with dried beans or rice. Guess which cylinder would hold more or whether they would hold the same amount. In other words, compare the volumes of the two cylinders. Explain your guess as well as you can.

2. Which of the two cylinders has more lateral surface area? Explain your answer.

Cereal Box Sizes

A standard box of Nutty Crunch cereal is 10 inches tall, 8 inches wide, and 2 inches deep.

1. Find the volume of the box.

2. The manufacturer is thinking about selling Nutty Crunch in other box sizes. Find the volume of each of these sizes.

   a. A minisize box that is half as tall, half as wide, and half as deep as the standard box

   b. An institutional-size box that is three times as tall, three times as wide, and three times as deep as the standard box

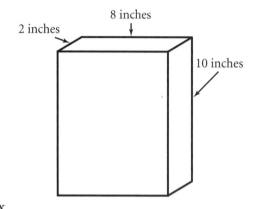

2 inches · 8 inches · 10 inches

3. Compare the volumes you found in Question 2 with the volume of the standard box. Look for more detail than simply, "The minisize is smaller. The institutional size is larger."

4. a. Find the volume of a supersize box—five times as tall, five times as wide, and five times as deep as the standard box. Try to do this without computing the dimensions of the box. Use your comparison in Question 3 for ideas.

   b. State a general principle for what happens to the volume of a box when each dimension is multiplied by the same amount.

   c. Explain your answer to part b. Use diagrams as needed.

More About Cereal Boxes

1. The manufacturer of Nutty Crunch wants to know how much cardboard would be used for the boxes described in *Cereal Box Sizes*. For the sake of simplicity, ignore the cardboard wasted in the manufacturing process. Also ignore places where the cardboard overlaps. In other words, simply find the total surface areas of the boxes.

   a. Find the surface area of the standard box.

   b. Find the surface area of the minisize box.

   c. Find the surface area of the institutional-size box.

2. Compare the three surface areas. As in *Cereal Box Sizes,* do more than simply say which is larger and which is smaller.

3. a. Find the surface area of the supersize box. Try to do this without computing the dimensions of this box.

   b. State a general principle for what happens to a box's surface area when each dimension is multiplied by the same amount.

   c. Explain your answer to part b. You might want to use diagrams.

A Size Summary

In *Cereal Box Sizes* and *More About Cereal Boxes,* you examined how the volume and surface area of a rectangular solid change if its dimensions are all multiplied by a specific scale factor.

1. Summarize your conclusions from those two activities.

2. Consider a rectangular solid that is 2 inches by 3 inches by 5 inches.

   a. Find the volume and surface area of this solid.

   b. Find the volume and surface area of a rectangular solid that is 6 inches by 9 inches by 15 inches using the conclusions you stated in Question 1. Explain how you used those conclusions.

   c. Find the dimensions of a rectangular solid with the same shape as the original but with a volume of 150 cubic inches. Give the dimensions to the nearest tenth of an inch.

   d. Find the dimensions of a rectangular solid with the same shape as the original but with a surface area of 124 square inches. Give the dimensions to the nearest tenth of an inch.

Back to the Bees

You will now combine all the ideas you've been studying with the concept of tessellation to solve the unit problem. You will then compile your portfolio and write your cover letter.

A-Tessellating We Go

The search for the best shape for a honeycomb cell has been considerably narrowed. You now want to find a shape that meets these conditions.

- The figure is a right prism.
- The base of the prism is a regular polygon.
- The base is a polygon that tessellates.

So you need to answer this question.

Which regular polygons tessellate?

Experiment! You can use pattern blocks for some regular polygons. You may want to cut out your own polygons to investigate the rest.

Once you decide which regular polygons tessellate, look for a way to prove that you have found them all.

A Portfolio of Formulas

In the course of solving the honeycomb problem of *Do Bees Build It Best?* you have developed many formulas related to area and volume. You will now collect these formulas to include them in your portfolio.

Look back over your work for this unit. Write down all the formulas you have used, with an explanation of each one. Be sure to define any variables, and include sketches to make the formulas clearer and more useful. Make sure your explanations will make sense to you when you look back at your portfolio at a future time.

Do Bees Build It Best? Portfolio

Now that *Do Bees Build It Best?* is completed, it is time to put together your portfolio for the unit. Follow these steps to compile your portfolio.

• Write a cover letter that summarizes the unit.

• Choose papers to include from your work in the unit.

• Discuss how your understanding of geometry has grown in this unit.

Cover Letter

Look back over *Do Bees Build It Best?* Describe the central problem and the main mathematical ideas in the unit. Your description should give an overview of how the key ideas of area and volume were developed and how they were used to solve the unit problem.

In compiling your portfolio, you will select some activities you think were important in developing the key ideas in this unit. Your cover letter should include an explanation of why you selected each item.

continued ◗

Selecting Papers

Your portfolio for *Do Bees Build It Best?* should contain these items.

- A Portfolio of Formulas
- A Problem of the Week
- Other key activities

  Identify two concepts in this unit for which you think your understanding improved significantly. For each concept, choose one or two activities that aided your understanding. Explain how each activity helped.

Personal Growth

Your cover letter for *Do Bees Build It Best?* should describe how the mathematical ideas developed in this unit. For the final part of your portfolio, write about your own personal development during this unit. You may want to address this question.

How do you think your understanding of geometry has grown?

Include any other thoughts you would like to share with a reader of your portfolio.

SUPPLEMENTAL ACTIVITIES

Measurement and the Pythagorean theorem are two of the main mathematical themes of this unit. They are reflected in these supplemental activities as well. Here are some examples.

- *Measurement Medley* is a series of activities on length, area, and volume.

- *Isosceles Pythagoras, Pythagorean Proof,* and *Pythagoras by Proportion* give you more opportunities to see and explain why the Pythagorean theorem is true.

- *Finding the Best Box* and *Another Best Box* follow up on the unit's opening-day activity. They ask you to examine the mathematics behind building the "best" box.

This section begins with a series of activities, *Memories of Year 1* and *More Memories of Year 1,* that look back at concepts from Year 1 of the IMP curriculum.

Memories of Year 1

Each of these activities builds on ideas from one of the Year 1 units of the IMP curriculum.

From *Patterns*

Consider the one-row In-Out table shown here.

In	Out
6	15

1. Find three different rules that fit this table.

2. Show that your three rules are really different (and not equivalent) by finding a single *In* value for which each rule gives a different *Out* value.

From *The Game of Pig*

Even-odd is a dice game. In each turn of the game, a player rolls a die—perhaps many times—and earns points according to these rules.

- Each time the player rolls the die:
    - If the die comes up even, the player adds the number on the die to the current point total.
    - If the die comes up odd, the player loses all points earned so far and gets a score of 0 for the turn.
- After each roll, the player has the option of taking the current total as the score for the turn or rolling again.

Suppose you are playing even-odd. Your current point total is 10, and you roll the die.

1. Find the probability of each possibility for your score after this roll.

2. Find the expected value of your score after this roll. That is, find the average score you would have in the long run if you rolled many times.

continued ◗

From *The Overland Trail*

One night along the trail, a heavy storm began just at midnight. When the storm started, George Whitman put the family's partly filled water barrel out to collect more rainwater. At 4:00 a.m., Clara Whitman woke up. She noticed that the water in the barrel was 165 millimeters deep. At 9:00 a.m., the depth was 180 millimeters.

1. Draw a graph showing the depth of the water as a function of the time elapsed since midnight. Assume the water level rose at a constant rate.

2. What was the depth of the water when George put out the water barrel?

3. By how much did the depth increase each hour?

4. Write a rule for the situation.

5. Suppose the rain continued at the same pace. When would the water level reach 200 millimeters?

From *The Pit and the Pendulum*

When Lion Forest plays golf, his tee shots go an average distance of 280 yards in the air, with a standard deviation of 20 yards. On a particular hole, Lion's shot needs to go at least 300 yards to make it over a lake. If the shot goes less than 240 yards, it will land in front of the lake. Then Lion can hit past the lake on his next shot. If the shot goes between 240 and 300 yards, it will either land in or roll into the lake.

continued ▶

1. What is the approximate probability that Lion's shot will make it past the lake?

2. What is the approximate probability that Lion's shot will avoid the lake?

From *Shadows*

A footlight at the front of the stage shines brightly on a performer. The performer casts a large shadow on the back wall of the stage.

The performer is 6 feet tall and stands 15 feet from the back wall. The shadow is 20 feet high.

How far is the performer from the footlight? Explain your reasoning.

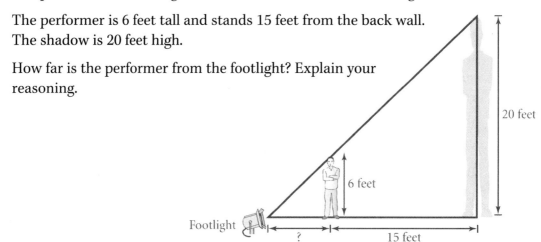

More Memories of Year 1

Each of these activities builds on ideas from one of the Year 1 units of the IMP curriculum.

From *Patterns*

1. For each table, write a statement in words that describes how to get the *Out* from the *In*.

    a.

In	Out
house	u
pickle	e
canary	a
baseball	e
flower	e

    b.

In	Out
sour	3
bitter	5
sweet	4
salty	4
mild	3

2. Complete each In-Out table based on the rule shown.

    a. $Out = 7 \cdot In - 5$

In	Out
0	
1	
2	
3	
4	

    b. $Out = 5 \cdot In + 9$

In	Out
0	
1	
2	
3	
4	

continued ▶

3. For each table, fill in the missing entry in the *Out* column with an algebraic expression using the variable given in the table.

a.

In	Out
0	4
1	7
2	10
3	13
4	16
t	

b.

In	Out
0	−4
1	1
2	6
3	11
4	16
m	

From *The Game of Pig*

A company that makes dice has made a huge error! They made a bunch of dice with the wrong dots. The center dot is missing from the five side of the dice, which makes it look like a four. All the other sides are the same, so each die shows 1, 2, 3, 4, 4, and 6.

Imagine playing the counters game with these dice. You will roll two dice and add the results.

1. Are there any numbers you can't roll with a pair of these dice that you could roll with a regular pair of dice? Are there any new numbers that you can roll?

2. Make an area model for the outcomes of rolling a pair of these dice.

3. Find the probability of rolling 8.

4. Find the probability of rolling an even number.

5. Find the probability of rolling an odd number.

6. Find the probability of rolling 13.

continued ⬦

From *The Overland Trail*

The Johnson and Pender families love to eat beans. Every day, they eat beans. Every day, they eat the same amount of beans. Running out of beans would be terrible for them. So they keep careful records of how many pounds of beans they have left.

The Johnson family begins with 100 pounds of beans and eats 5 pounds a day. The Pender family begins with 140 pounds of beans and eats 10 pounds a day.

1. On the same set of axes, graph the amount of beans each family has left for the next 10 days.

2. Is there a day when the Johnson family will have the same amount of beans as the Pender family? If so, show where this is on your graph.

3. Write a rule for the number of pounds of beans the Johnsons have after D days.

4. Write a rule for the number of pounds of beans the Penders have after D days.

From *The Pit and the Pendulum*

The Bayside Bolt Company makes big boxes of bolts. A sample of boxes contains an average of 550 bolts with a standard deviation of 20 bolts.

1. Draw a normal distribution and mark the given information.

2. Suppose you buy a box of bolts from Bayside Bolt Company. What is the probability that the box will contain fewer than 550 bolts?

3. Complete this statement by filling in the blanks.

    If you pick a box at random, 68% of the time you should get a box with between _____ and _____ bolts.

4. What is the probability of getting a box with more than 590 bolts?

continued ▶

From *Shadows*

1. In this diagram, the woman is 5 feet tall and casts a 4-foot shadow. The tree casts a 16-foot shadow. How tall is the tree?

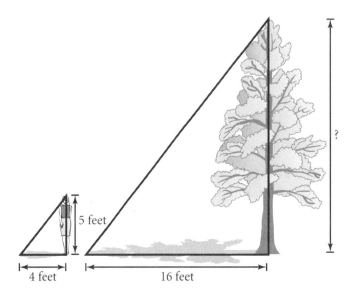

5 feet

4 feet   16 feet

2. These trapezoids are similar. Find the values of *x* and *y*.

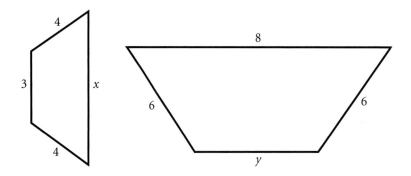

4

3   *x*

4

8

6   6

y

Measurement Medley

Length

Measurement plays an important role in this unit, and length is one of the fundamental aspects of measurement.

1. Put a meterstick across the room where you can see it but not reach it. Then use visual estimation to cut pieces of string with each of these lengths.

   - 10 centimeters
   - 38 centimeters
   - 2 meters

   Measure your pieces of string. Compare the results with what you were trying to get.

2. Use string and a globe to estimate each distance below in kilometers. To do this, you will need to know that the distance around the earth at the equator is about 40,000 kilometers.

   - From San Francisco, California, to New York, New York
   - From Hanoi, Vietnam, to Atlanta, Georgia
   - From Capetown, South Africa, to London, England
   - Two locations of your choice

continued ▶

Area

Area is generally measured in square units of some kind. In this part of the activity, you'll work with square centimeters, square kilometers, and square inches.

1. Use centimeter graph paper, a globe, and the fact that the distance around the earth at the equator is about 40,000 kilometers to estimate the area of the continental United States in square kilometers.

2. Suppose that 0.1 ounce of a certain skin cream will cover 50 square inches of skin. Estimate how much cream you would need to cover your arm.

3. Use centimeter graph paper for this question.
   a. Draw at least five different rectangles, each with a perimeter of 30 centimeters.
   b. Which of your rectangles in part a has the smallest area? The largest area?
   c. Draw at least five different rectangles, each with an area of 24 square centimeters.
   d. Which of your rectangles in part c has the smallest perimeter? The largest perimeter?
   e. What conclusions can you draw from your work on this question?

Volume

Length, area, and then comes volume—in this part of the activity, you will explore this third dimension of measurement.

1. Use centimeter cubes and a globe to estimate the volume of the earth in cubic kilometers. Use the fact that the distance around the earth at the equator is about 40,000 kilometers.

continued ▶

2. Find some objects around your home that have approximately these volumes.

   a. 1 cubic inch

   b. 1 cubic foot

   c. 1 cubic meter

3. Fill a jar with marbles and then estimate the fraction of the jar's volume that is air. Do you think this fraction depends on the shape of the jar? Explain.

4. Estimate the volume, in cubic centimeters, of each of these objects.

   a. A penny

   b. A pencil

   c. A rock (Find a rock to use for this.)

How Many of Each?

In this activity, you will use the triangle and the square pattern blocks as units to measure area. Then you will compare the values you get from using different units.

1. Build a shape out of square blocks.

    a. Record the area of your shape using the square as the unit.

    b. Find the approximate area of your shape using the triangle as the unit.

2. Build a second shape, using twice as many squares as you used in Question 1.

    a. Record the area of your new shape using the square as the unit.

    b. Find the approximate area of the new shape using the triangle as the unit.

3. a. Draw some figures on paper and find their areas. First use the square as the unit, and then use the triangle as the unit.

    b. Record your results (including those from Questions 1 and 2) in an In-Out table. Use the area measured in squares as the *In* and the area measured in triangles as the *Out*.

    c. Look for an approximate rule for your table.

All Units Are Not Equal

For this activity, you will need two shapes of different sizes to use as units of area. If you like, you can simply cut two irregular shapes out of heavy paper.

1. Using the smaller shape as the unit of area, estimate the areas of two flat objects around your house. Record those results.

2. Estimate the same areas using your larger shape as the unit. Record those results.

3. Describe how you estimated the areas. Explain any differences between finding your estimates using the larger unit and finding your estimates using the smaller unit.

4. Imagine that you measured a surface using your smaller shape as the unit and got 5 as the value of the area. What would you get as the area if you used your larger shape as the unit? Explain your reasoning.

Geoboard Squares

In *Checkerboard Squares* (a Problem of the Week from the Year 1 unit *Patterns*), you were asked to find the number of squares (of any size) on an 8-by-8 checkerboard. In that problem, the squares all had vertical and horizontal sides because they were all made of one or more individual colored squares from the checkerboard.

The geoboard, however, has other kinds of squares. For example, this geoboard shows a square with slanted sides.

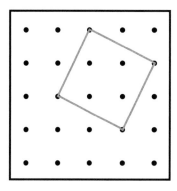

Your task is to find how many squares of all kinds are on a geoboard. The only condition is that the vertices of the squares must be at pegs. Don't simply find out how many *kinds* of squares there are. Consider every square on the geoboard as an individual example. (Not every four-sided figure is a square.) Be sure all the figures you include are squares.

Start with the standard 5-peg-by-5-peg geoboard, and then consider geoboards of other sizes. Look for a method for calculating the number of squares on an *n*-peg-by-*n*-peg geoboard.

Extra challenge: Generalize your results to an *m*-peg-by-*n*-peg rectangular geoboard.

More Ways to Halve

In the activity *Halving Your Way,* you looked at ways to divide the rectangle shown below into two parts with equal area. You had to do this with a single rubber band attached at pegs on or inside the rectangle.

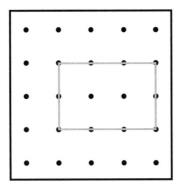

In this activity, you can attach the rubber band to *any* pegs on the geoboard. For example, the diagram below shows one new method you can use. How many *different* ways can you find now to divide the rectangle into two equal parts?

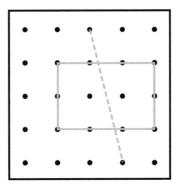

Isosceles Pythagoras

The Pythagorean theorem says,

> When a triangle has a right angle, the sum of the areas of the squares built on the two legs equals the area of the square built on the hypotenuse.

Your task is to find a simple proof of this statement for the special case of an isosceles right triangle.

Consider right triangle ACB shown here, where $AC = BC$. Prove that the sum of the areas of the two smaller shaded squares is equal to the area of the larger shaded square. Make your proof as simple as possible.

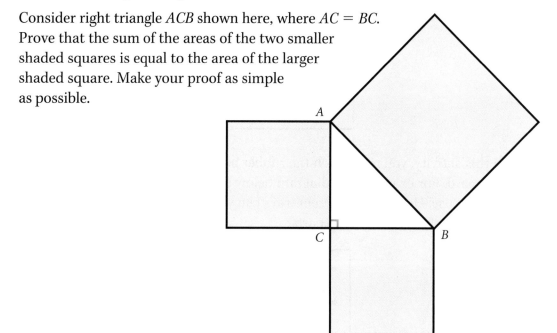

More About Pythagorean Rugs

This diagram was used in the activity *Proof by Rugs* to prove the Pythagorean theorem.

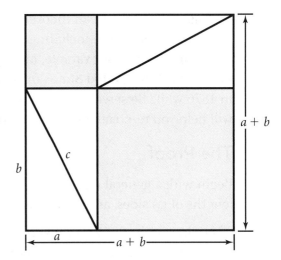

As you saw in that activity, this diagram starts with a right triangle with legs of lengths a and b and a hypotenuse of length c. Four copies of that triangle are then placed inside a square of side length $a + b$, as shown.

In *Proof by Rugs,* you compared this diagram to a similar one and showed that the two diagrams had the same amount of shaded area. Your task now is to compare the shaded area in the diagram to the unshaded area in the same diagram.

Which is greater—the shaded area or the unshaded area? Does it depend on which triangle you start with? Are there cases where the shaded area and the unshaded area are equal? Prove your answers.

Pythagorean Proof

Because the Pythagorean theorem is so important, many people have developed proofs of it, including people who were not professional mathematicians. For example, James Garfield, who was elected president of the United States in 1880, is credited with creating a proof in 1876 while he served in the House of Representatives. This activity will help you re-create one of the better-known proofs.

The Proof

Begin with a general right triangle *ABC*. Use *a*, *b*, and *c* to represent the lengths of its sides, as shown below.

Then draw altitude $\overline{CD}$ from vertex *C* to the hypotenuse, as shown at the right. This creates two smaller right triangles, *ACD* and *CBD*, in addition to the original right triangle.

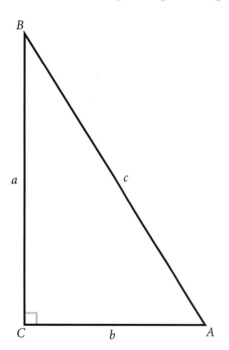
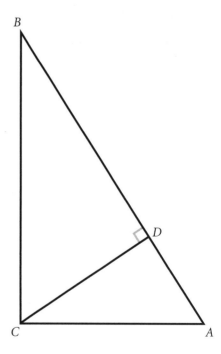

continued ▶

The proof is based on the fact that the three right triangles are all similar.

1. Prove that the three triangles *ABC*, *ACD*, and *CBD* are all similar.

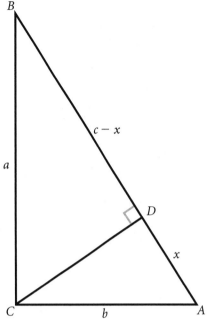

The next step of the proof is to set up some proportions based on similarity. To get started, look at $\overline{AB}$, the hypotenuse of the large triangle. Notice that it is broken into two parts by point *D*. Use *x* to represent the length of the segment from *A* to *D*. Use $c - x$ to represent the length of the segment from *D* to *B*, as shown here.

Based on this labeling, move on to Question 2.

2. Use the fact that the three triangles are similar to write several equations involving *a*, *b*, *c*, and *x*. These equations should state that certain ratios of sides are equal. Here are some hints.

   • Draw and label diagrams that show each triangle separately. You may want to turn or flip the triangles so it's clear how the sides match up.

   • Compare each of the two smaller triangles to the large one.

3. Work with the equations you developed in Question 2 to show that $a^2 + b^2 = c^2$.

Pythagoras by Proportion

Here is an outline of another proof of the Pythagorean theorem. This proof is based on the principle that the areas of similar polygons are proportional to the squares of corresponding sides.

In this diagram, the altitude from point C to segment AB divides the large triangle into two smaller triangles, labeled I and II. The legs of triangle I are labeled r and s. The angle at A is labeled θ. Consider the large triangle ABC as triangle III.

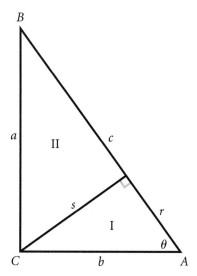

1. Express r and s in terms of b and θ. Show that the area of triangle I is equal to $b^2 \cdot \frac{1}{2} \cdot \cos \theta \cdot \sin \theta$.

2. Show that the area of triangle II is equal to $a^2 \cdot \frac{1}{2} \cdot \cos \theta \cdot \sin \theta$. Think about where θ appears in triangle II, and why.

3. Show that the area of triangle III is equal to $c^2 \cdot \frac{1}{2} \cdot \cos \theta \cdot \sin \theta$.

4. Use your results from Questions 1, 2, and 3 to show that $a^2 + b^2 = c^2$.

5. How does this proof use the principle that the areas of similar polygons are proportional to the squares of corresponding side lengths?

All About Escher

M. C. Escher (1898–1972) was a Dutch artist known for his imaginative use of geometry. This illustration is one of his creations.

Write a report about M. C. Escher, his life, and his work. Find out where he got some of his design ideas.

The Design of Culture

Geometric patterns, such as tessellations, have been used in the crafts and art of many civilizations as far back as recorded history goes.

Choose a culture—contemporary or ancient—and describe the use of geometry in some aspect of that culture's artistry. For instance, you might look at weavings, architecture, or fashion.

Close-up of a Celtic ornamental design from "Celtic Designs," copyright © 1981 by Rebecca McKillip, used by permission of Stemmer House Publishers.

Drawing from a Hmong textile created by hill tribe people in Southeast Asia from "Southeast Asian Designs," copyright © 1981 by Caren Caraway, used by permission of Stemmer House Publishers.

Tessellation Variations

In the POW *Tessellation Pictures,* you cut and rearranged the pieces of a rectangular index card to make a new shape. You then put many copies of your shape together to form a tessellation drawing like the one shown here.

Your task in this activity is to create some new tessellation drawings. This time start with tessellating figures that are not rectangles. You might try triangles and parallelograms, for example.

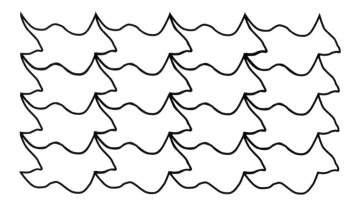

Beyond Pythagoras

The Pythagorean theorem states a relationship among the lengths of the sides of a right triangle. But what about triangles in general?

Consider the general triangle shown here, with sides of lengths a, b, and c. Is there any relationship among those lengths when $\angle C$ is not a right angle?

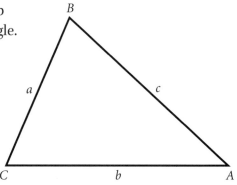

The answer, you'll be glad to hear, is yes. In fact, by drawing just one extra line and using some algebra, you can see what that relationship is, at least for the case where $\angle C$ is an acute angle. (You may see a connection between the method described here and the activity *Leslie's Fertile Flowers*.)

In this new diagram, the altitude has been drawn from vertex B to point D. Several lengths have been labeled as well.

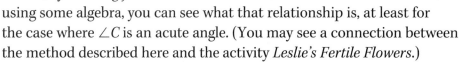

1. Apply the Pythagorean theorem to triangles *ABD* and *BCD* to get two expressions for h^2. One expression should be in terms of a and x. The other should be in terms of b, c, and x.

2. Set the two expressions for h^2 equal to each other. Simplify to get an expression for c^2 in terms of a, b, and x.

Note: If $\angle C$ were 90°, then x would be 0. Replacing x with 0 in your result from Question 2 should give you the Pythagorean theorem.

3. Use trigonometry to replace x with an expression in terms of a and $\angle C$.

The equation that results from Question 3 is called *the law of cosines*.

4. How could you use the equation from Question 2 to find x and answer the question in *Leslie's Fertile Flowers*?

5. *Extra challenge*: How would you have to adjust your work if $\angle C$ were an obtuse angle?

Comparing Sines

In the activity *Leslie's Fertile Flowers*, you saw that in a diagram like this, you can use the Pythagorean theorem to get two different equations involving h. You could then compare the two expressions to get more information.

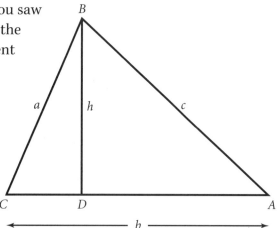

Now you will use trigonometry to get two different equations involving h. You will use the results to get a relationship involving the side lengths and the angles of the triangle.

1. Use trigonometry in right triangle *BCD* to write an expression for h in terms of a and $\angle C$.

2. Get a similar equation for h using triangle *BAD*.

3. Combine your two results to get an equation involving just a, c, and the two angles A and C. Write this equation using just a and A on one side and just c and C on the other side.

4. Describe how you would use a different altitude to show that an expression exists involving just b and $\angle B$ that is equal to your expression with just a and A from Question 3.

The statement that the three expressions from Questions 2 and 3 are equal is called *the law of sines*.

5. *Extra challenge:* How would you have to adjust your work if one of the angles were obtuse?

Hero and His Formula

In *Leslie's Fertile Flowers,* you found the area of a triangle when your only information was the lengths of the three sides. The method you used can be applied to any triangle. In fact, the formula known as *Hero's formula* expresses the area of a triangle in terms of the lengths of its sides.

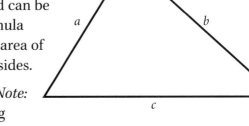

Your task is to prove Hero's formula. (*Note:* Questions 2 and 3 are about simplifying the formula shown in Question 1. Work on Questions 2 and 3 even if you can't prove the formula in Question 1.)

1. Suppose you have a triangle with sides of lengths a, b, and c, as shown above. Use the method from *Leslie's Fertile Flowers* to show that the area of this triangle is given by the equation

$$A = \frac{\sqrt{2a^2b^2 + 2a^2c^2 + 2b^2c^2 - a^4 - b^4 - c^4}}{4}$$

Reminder: The method in *Leslie's Fertile Flowers* has these five steps.

- Draw a diagram like this one in which the altitude is shown and the base is split into two parts of lengths x and $c - x$.

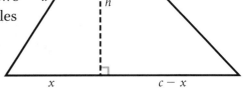

- Use the Pythagorean theorem to write two different equations involving the variables in the diagram.

- Combine the two equations to get an equation that does not involve h. Solve this equation for x.

- Find the height of the triangle.

- Find the area of the triangle.

When you solve for x (in the third step), you will get an expression involving the lengths a, b, and c. Finding h once you have x is messy but doesn't involve anything more than the Pythagorean theorem and some algebra.

continued ▶

2. Using algebra, show that the equation below is **equivalent** to the equation in Question 1.

$$A = \tfrac{1}{4}\sqrt{(a+b+c)(a+b-c)(b+c-a)(a+c-b)}$$

3. The *semiperimeter* of a triangle is defined as half the triangle's perimeter. Hero's formula can be simplified by letting the variable s stand for the semiperimeter. In other words, s is defined by the equation

$$s = \frac{a+b+c}{2}$$

Using this definition for s, simplify the result from Question 2 to give the equation

$$A = \sqrt{s(s-a)(s-b)(s-c)}$$

This simplification is the usual form of Hero's formula.

Toasts of the Round Table

When someone proposes a toast, people often clink their glasses together in celebration. Toasting wasn't always done this way. When King Arthur's knights toasted, each knight tapped his lance against the lance of another knight. If possible, every knight tapped lances with every other knight at the table.

As you may know, King Arthur's knights were usually seated at a round table. They didn't want to leave their seats, and the table was large. So knights opposite each other couldn't always reach far enough to tap lances.

Imagine 30 knights spaced equally around a round table. Each knight can extend his lance to reach up to 10 feet from his sitting position. (That takes arm length into account.) The table has a radius of 12 feet.

King Arthur walks into the room and raises his glass. Standing near the table, he drinks in honor of his knights. In response, each knight taps lances with all other knights that are close enough.

1. How many lance taps are there? Explain your answer fully.

2. Consider variations on this situation. Here are some things you might change.

   - The number of knights
   - The radius of the table
   - The distance each knight can reach

   Consider changes such as these, and generalize your discoveries as much as you can.

What's the Answer Worth?

Sometimes we measure or estimate something and then make calculations based on those estimates. When your original numbers are only approximations, knowing the accuracy of your final answer is important.

1. Two students measured the triangle shown here. They found its base to be 4.8 centimeters and its altitude to be 3.4 centimeters. Both of these measurements were found to the nearest tenth of a centimeter.

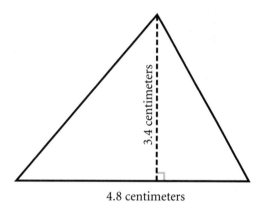

3.4 centimeters

4.8 centimeters

a. Think about the fact that the measurements are rounded to the nearest tenth. What is the *maximum* length the base could be? What is the *maximum* length the altitude could be?

b. Use your answers to determine the triangle's maximum possible area.

c. What is the minimum the area could be?

2. Suppose you are taking a test that shows the figure above with no explanation. You are asked to find the area of the triangle. How would you answer the question?

Finding the Best Box

In the activity *Building the Biggest,* you looked at how to build a box with the greatest possible volume from a sheet of construction paper. You were asked to build a box with four sides and a bottom but no top. Now you will return to that task with a few new details thrown in.

One way to build such a box from a rectangular sheet of paper is to begin by cutting out squares at each of the four corners of the sheet. The cuts are shown with the dashed lines in the diagram at the right.

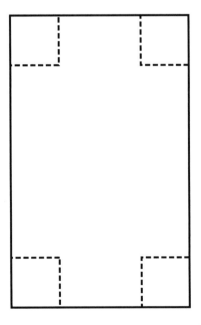

continued ▶

The shape that remains after the corners are cut out is shown at the right. This remaining piece of paper can then be folded along the dashed lines, with the flaps lifted together to form the sides of the box.

The central rectangular area (within the dashed lines) becomes the bottom of the box.

You will focus on what happens if you use this method starting with a *square* sheet of paper.

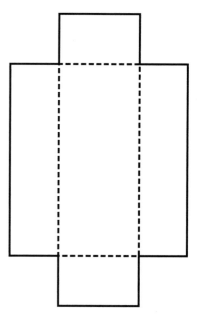

1. Start with a sheet of paper 12 inches by 12 inches. Use x to represent the side of each cut-out corner.

   a. Find an expression in terms of x for the volume of the resulting box.

   b. Determine what value of x will maximize this volume.

   c. What is that maximum volume?

2. Repeat Questions 1a and 1b, but start with square sheets of paper of different sizes.

3. Describe any patterns you find for how the size of the corner of the "best" box depends on the size of the square sheet of paper.

Another Best Box

Recycling can change the meaning of "best."

Finding the Best Box asks what happens if you build an open-top box by cutting out square corners from a square sheet of paper and folding up the sides. Specifically, Question 1 asks for the maximum possible volume if the sheet of paper is 12 inches by 12 inches.

In that activity, the cut-out corners are basically wasted. Presumably you could get more volume out of the same sheet of paper if you could somehow use the surface area of those corners. Suppose you have the full 144 square inches available to build an open-top box. In other words, you want to build an open-top box whose total surface area—bottom and four sides—is the same as the total area of a 12-inch-square sheet of paper. Assume the base of your box still has to be a square.

1. Begin with a tall, thin box whose base is a 2-inch square.
    a. Figure out how tall the box should be to get a total surface area of 144 square inches. (*Remember:* You are counting the bottom and four sides, but no top.)
    b. Find the volume of that box.

2. Repeat Question 1 for a short, fat box. You pick the size of the square base.

3. To generalize, suppose the box's base is a square that is *b* inches on each side. Answer both parts of Question 1 for this situation. Your answers will be expressions in terms of *b*.

4. Find the maximum possible volume and the value of *b* that gives the volume you found in Question 3. You may need to approximate these answers. You may also want to use a graphing calculator.

From Polygons to Prisms

In the activity *Back on the Farm,* you helped farmer Minh minimize the amount of paint he needed for his barn. He is now focusing on maximizing the volume of his barn.

Farmer Minh still wants the barn to be in the form of a prism, with vertical sides and a base that is a regular polygon. He has decided that he can afford to buy 3000 square feet of board to use for the lateral sides. He also wants the barn to be 10 feet tall.

What shape should farmer Minh use for the base of the prism to maximize the volume of his barn? Explain your reasoning.

What Else Tessellates?

In the activity *A-Tessellating We Go,* you investigated which regular polygons can be used to tessellate the plane. This diagram shows how equilateral triangles can be used to do this.

Now you will consider nonregular polygons. Start with triangles, and investigate which tessellate and which do not. Then move on to quadrilaterals. See what you can find out, and write a report summarizing your findings.

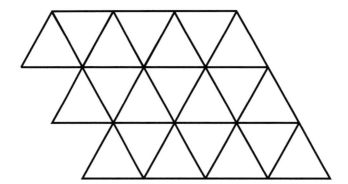

Cookies

Systems of Equations and Linear Programming

Cookies—Systems of Equations and Linear Programming

Cookies and Inequalities

The central problem of this unit involves helping a bakery to maximize its profits. The problem is complex. First you will organize all of the information and express the bakery's situation in algebraic terms, using linear inequalities and linear expressions.

Sonia Mena creates a graph to help solve the bakery problem.

How Many of Each Kind?

Abby and Bing Woo own a small bakery that specializes in cookies. They make only two kinds of cookies—plain and iced. They need to decide *how many dozens* of each kind of cookie to make for tomorrow.

The Woos know that each dozen *plain* cookies requires 1 pound of cookie dough (and no icing). Each dozen *iced* cookies requires 0.7 pound of cookie dough and 0.4 pound of icing. They also know that each dozen plain cookies requires about 0.1 hour of preparation time. Each dozen iced cookies requires about 0.15 hour of preparation time. Finally, they know that no matter how many of each kind they make, they will sell them all.

Three factors limit the Woos' decision.

- The ingredients available: They have 110 pounds of cookie dough and 32 pounds of icing.

- The oven space available: They have room to bake a total of 140 dozen cookies.

- The preparation time available: Together they have 15 hours for cookie preparation.

continued ▶

Why on earth should the Woos care how many cookies of each kind they make? Well, you guessed it! They want to make as much profit as possible. Plain cookies sell for $6.00 a dozen and cost $4.50 a dozen to make. Iced cookies sell for $7.00 a dozen and cost $5.00 a dozen to make.

The big question is

How many dozens of each kind of cookie should Abby and Bing make so their profit is as high as possible?

1. Begin to answer the big question.
   a. Find one combination of dozens of plain cookies and dozens of iced cookies that will satisfy all of the conditions in the problem.
   b. Figure out how much profit the Woos will make on that combination of cookies.

2. Now find a different combination of dozens of cookies that fits the conditions but yields a greater profit.

Adapted from *Introduction to Linear Programming*, 2nd ed., by R. Stansbury Stockton (Boston: Allyn and Bacon, 1963).

A Simpler Cookie

The Woos have a rather complicated problem to solve. Let's make it simpler. Finding a solution to a simpler problem may lead to a method for solving the original problem.

Assume the Woos still make both plain and iced cookies and have 15 hours for cookie preparation. But now assume they have unlimited amounts of cookie dough and icing. Also assume that they have an unlimited amount of oven space.

The other information is unchanged.

- Preparing a dozen plain cookies requires 0.1 hour.
- Preparing a dozen iced cookies requires 0.15 hour.
- Plain cookies sell for $6.00 a dozen.
- Plain cookies cost $4.50 a dozen to make.
- Iced cookies sell for $7.00 a dozen.
- Iced cookies cost $5.00 a dozen to make.

As before, the Woos know that no matter how many of each kind of cookie they make, they will sell them all.

1. Find at least five combinations of plain and iced cookies that the Woos could make without working more than 15 hours. For each combination, compute the profit.

2. Find the combination of plain and iced cookies that you think would give the Woos the most profit. Explain why you think no other combination will yield a greater profit.

Investigating Inequalities

Part I: Manipulating Inequalities

You have learned about ways to change equations so they still hold true. For instance, suppose you have a true equation—that is, two expressions that are equal. You could add the same quantity to both sides of the equation, and the resulting expressions would still be equal.

For example, the statement $3 + 8 = 5 + 6$ is true, because $3 + 8$ and $5 + 6$ are both equal to 11. If you add 7 to both sides, the resulting statement is $3 + 8 + 7 = 5 + 6 + 7$. This statement is also true.

1. Investigate whether similar principles hold true for inequalities. The inequality $4 > 3$ is true. Starting with this inequality, perform each of these operations and examine whether the resulting statement is true.

   - Add the same number to both sides of the inequality.
   - Subtract the same number from both sides of the inequality.
   - Multiply both sides of the inequality by the same number.
   - Divide both sides of the inequality by the same number.

   For example, if you multiply both sides of the inequality $4 > 3$ by 2, the statement becomes $4 \cdot 2 > 3 \cdot 2$. Your task for each operation is to determine whether the new statement is true no matter what "the same number" is.

   Try different possibilities for "the same number." Use both positive and negative values.

2. After you finish investigating the inequality $4 > 3$, try a different true inequality. See whether you reach the same conclusions.

3. When you are done exploring, state your conclusions. Make them as general as possible.

continued ▶

Part II: Graphing Inequalities

If an inequality contains a single variable, you can picture all the numbers that make the inequality true by shading them on a number line. This is called the graph of the **inequality.** An inequality using $<$ or $>$ is called a *strict inequality.* An inequality using $\leq$ or $\geq$ is called a *nonstrict inequality.*

For example, the colored portion of this number line represents the graph of the strict inequality $x < 4$.

The open circle at the number 4 on the number line means that the number 4 is not included in the graph. (The number 4 is not included because substituting 4 for x gives a false statement.)

If you want to include a particular number as part of the graph, you mark that point with a filled-in circle. For example, the colored portion of the next diagram represents the graph of the nonstrict inequality $x \leq 4$.

4. Draw the graph of the inequality $x > -2$.

5. Draw the graph of the inequality $x \leq 0$.

6. What inequality goes with this graph?

7. How would you use inequalities to describe this graph?

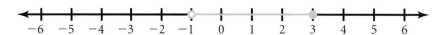

My Simplest Inequality

In *Investigating Inequalities,* you started with a true inequality. You explored which operations you could perform on both sides to create another true inequality.

When inequalities involve variables, you want to know if the operation produces an **equivalent inequality.** As with equations, two inequalities are equivalent if any number that makes one of them true also makes the other true.

For example, the inequalities $x + 2 < 9$ and $2x + 4 < 18$ are equivalent. The numbers that make both inequalities true are precisely the numbers less than 7. For instance, substituting 5 for x makes both statements true. Substituting 10 for x makes both false. That is, $5 + 2 < 9$ and $2 \cdot 5 + 4 < 18$ are true, while $10 + 2 < 9$ and $2 \cdot 10 + 4 < 18$ are false.

Part I: One Variable Only

If an inequality has only one variable, you can often find an equivalent inequality that essentially gives the solution. For instance, by subtracting 2 from both sides of $x + 2 < 9$, you get the equivalent inequality $x < 7$. This tells you that the solutions to $x + 2 < 9$ are the numbers less than 7 (and only those numbers).

continued ▶

1. For each inequality, perform operations to get equivalent inequalities until you obtain one that shows the solution.

   a. $2x + 5 < 8$

   b. $3x - 2 \geq x + 1$

   c. $3x + 7 \leq 5x - 9$

   d. $4 - 2x > 7 + x$

Part II: Two or More Variables

When an inequality has more than one variable, you can't put it into a form that directly describes the solution. But you can often write the inequality in a simpler, equivalent form by combining terms.

For example, suppose you start with the inequality

$$9x - 4y - 2 \geq 3x + 10y + 6$$

These steps will produce a sequence of simpler, equivalent inequalities.

$9x - 2 \geq 3x + 14y + 6$ (adding $4y$ to both sides)

$6x - 2 \geq 14y + 6$ (subtracting $3x$ from both sides)

$6x \geq 14y + 8$ (adding 2 to both sides)

All the **coefficients** in $6x \geq 14y + 8$ are even. So, you can divide both sides of the inequality by 2. This gives $3x \geq 7y + 4$.

2. Each inequality in the sequence is equivalent to the original inequality. However, $3x \geq 7y + 4$ seems to be the simplest of all.

   a. Find numbers for x and y that fit the inequality $3x \geq 7y + 4$.

   b. Substitute the numbers that you found in part a into the original inequality, $9x - 4y - 2 \geq 3x + 10y + 6$. Verify that these numbers make the inequality true.

   c. Find numbers for x and y that do not fit the inequality $3x \geq 7y + 4$.

continued ▶

d. Substitute the numbers you found in part c into the original inequality, $9x - 4y - 2 \geq 3x + 10y + 6$. Verify that these numbers make the inequality false.

e. Explain why your work in parts a to d is not enough to prove that the two inequalities are equivalent.

3. For each inequality, perform the appropriate operations to get simpler, equivalent inequalities.

a. $x + 2y > 3x + y + 2$

b. $\frac{x}{2} - y \leq 3x + 1$

c. $0.2y + 1.4x < 10$

Simplifying Cookies

As you have seen, you can express the **constraints** in the unit problem as inequalities using two variables. Suppose you use P to represent the number of dozens of plain cookies and I to represent the number of dozens of iced cookies. One way to write these inequalities is

$$P + 0.7I \leq 110 \qquad \text{(for the amount of cookie dough)}$$

$$0.4I \leq 32 \qquad \text{(for the amount of icing)}$$

$$P + I \leq 140 \qquad \text{(for the amount of oven space)}$$

$$0.1P + 0.15I \leq 15 \qquad \text{(for the amount of preparation time)}$$

1. Find at least one equivalent inequality for each of these "cookie inequalities." If possible, find one that you think is simpler than the given inequality.

2. For each of the original inequalities, do these things.
   a. Find a number pair for P and I that fits the inequality. Also find a number pair that does not fit the inequality.
   b. Verify that the number pair that fits the inequality also fits any equivalent inequalities you found.
   c. Verify that the number pair that does not fit the inequality also does not fit any of the equivalent inequalities you found.

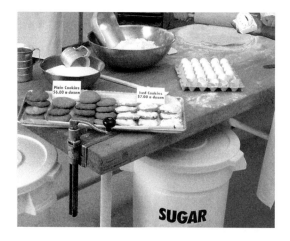

Picturing Cookies

You have turned the bakery's problem into a set of inequalities and a profit expression. That's a good first step toward understanding the problem. Over the next several days, you will examine how to represent these inequalities using graphs. The graphs will give you, at a glance, a picture of the Woos' various options.

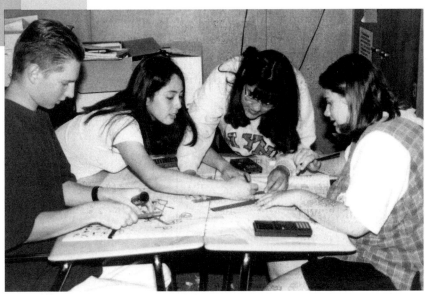

Mark Hansen, Jennifer Rodriguez, Karla Viramontes, and Robin LeFevre make a group graph for a cookie inequality.

Picturing Cookies—Part I

Using graphs, you can turn symbolic relationships into geometric ones. Geometric relationships are visual, so they are often easier to think about than algebraic statements.

One of the constraints in *How Many of Each Kind?* is based on oven-space limitations. The Woos can make at most 140 dozen cookies. You can represent this constraint symbolically by the inequality

$$P + I \leq 140$$

where P is the number of dozens of plain cookies and I is the number of dozens of iced cookies.

Choose a color to use for combinations of plain and iced cookies that satisfy the constraint—that is, combinations that total 140 dozen cookies or fewer. Choose another color for combinations that do not satisfy the constraint—that is, combinations that total more than 140 dozen cookies.

Some Examples

What color would you use for the point (20, 50)? In other words, does the combination of 20 dozen plain cookies and 50 dozen iced cookies fit the constraint? You can check by substituting 20 for P and 50 for I

continued ▸

in the inequality $P + I \leq 140$. Because $20 + 50 \leq 140$ is a true statement, you would use the first color for the point (20, 50).

What about 90 dozen plain cookies and 120 dozen iced cookies? This combination does not satisfy the constraint because the statement $90 + 120 \leq 140$ is not true. You would use the second color for the point (90, 120).

Your Task

Your task is to plot points of both types. Then you will describe the graph of the inequality itself. The graph of the inequality consists of all points that fit the constraint—that is, all points of the first color.

Steps 1 to 3 in Question 1 explain what you need to do. Draw your final diagram on a sheet of grid chart paper. If you have time answer Question 2, which deals with other constraints.

1. Do these steps for the oven-space inequality.

   Step 1: Each group member will test many pairs of numbers to see which ones satisfy the constraint. On one set of coordinate axes, group members will plot all of their number pairs using the appropriate colors.

   Step 2: Make sure your group has many points of both colors. After some experimentation, you may need to change the scales on your axes to show both types of points. If necessary, redraw your axes with new scales and replot the points you have already found.

   Step 3: Continue adding points of each type in the appropriate color. Continue until you see the "big picture." In other words, continue until you are sure what the overall diagram will look like. With your final diagram include a statement describing the graph of the inequality itself (the points of the first color). Also explain why you think your description is correct.

2. Graph each of the remaining constraints on its own set of axes. Follow the process described in Question 1, Steps 1 to 3 above, or use what you learned in Question 1 about the "big picture."

Inequality Stories

You have seen that you can use inequalities to describe real-world situations.

For example, in *How Many of Each Kind?* each dozen plain cookies uses 1 pound of cookie dough and each dozen iced cookies uses 0.7 pound of cookie dough. However, the Woos have only 110 pounds of cookie dough. This limitation can be described by the inequality $P + 0.7I \leq 110$. In this inequality, P is the number of dozens of plain cookies and I is the number of dozens of iced cookies.

In this activity, you will look further at the relationship between real-world situations and inequalities.

Part I: Stories to Inequalities

Use variables to write an inequality that describes each situation. Explain what your variables represent.

1. Rancher Gonzales has built the corral for her horses. Now she's building a pen for her pigs. As she isn't worried about efficiency for the pigs, she decides to use a boring old rectangle. The pigs need an area at least 150 square meters. Rancher Gonzales has to decide what dimensions to make the pen.

2. Al and Betty want to buy a really fancy spinner that costs $200. They each have some money of their own. Al's parents will contribute $2 for every $1 that Al spends. Betty's grandmother will match Betty's contribution exactly. But even if Al and Betty combine all their own money with these additional funds, they still won't have enough.

Part II: Inequalities to Stories

Create a real-world situation that each inequality might describe. Remember to explain what your variables represent.

3. $r < t + 2$ 4. $a + b + c \leq 30$ 5. $x^2 + y^2 \geq 81$

A Hat of a Different Color

Once upon a time, many years ago and very far away, there lived a wise high school teacher whose students always complained noisily that they had too much homework (and too many POWs).

The wise teacher offered the three noisiest students a deal. He showed them that he had two red hats and three blue hats. The deal worked like this.

> The three students would close their eyes. While their eyes were closed, the teacher would put a hat on each one's head. He would hide the other two hats.

> One at a time, the students would open their eyes. Each student would look at the hats on the other two students and try to determine which color hat was on his or her own head. At a given student's turn, that student could either guess what color hat he or she had on or pass.

> While the first student was deciding what to do, the other two students would keep their eyes closed.

> Once the first student either guessed or passed, the second student could open his or her eyes and either guess or pass. The third student kept his or her eyes shut.

> When the second student was finished, the third student could open his or her eyes and either guess or pass.

Any student who guessed correctly would not have to do any POWs for the rest of the semester. But any student who guessed wrong would not only have to do the POWs but also help grade everyone else's work. If a student decided to pass, his or her workload did not change.

continued

The students drew numbers to see who would go first. Then they closed their eyes. The wise teacher put a hat on each one's head and hid the remaining two hats.

Arturo, who was first, opened his eyes. He looked at the other students' hats and said, "Pass." He couldn't tell for sure, and he didn't want to guess in case he was wrong.

Next, Belicia opened her eyes and looked at the other students. She considered the fact that Arturo couldn't tell. Then she said, "Pass." She couldn't tell for sure either.

Carletta was third. She simply sat with her eyes still closed tightly and a big grin on her face. "I know what color hat I have on," she said. And she gave the right answer.

Your task is to figure out what color hat Carletta was wearing and how she knew for sure. The most important part of your POW write-up is your explanation of how she knew.

Reminder: Carletta didn't even look! You should also know that all three students were extremely smart. If they could have figured it out, they would have.

○ *Write-up*

1. *Problem Statement*

2. *Process*

3. *Solution:* Explain how you know what color hat Carletta has on.

4. *Evaluation*

5. *Self-assessment*

Healthy Animals

Curtis is concerned about his pet's diet. A nutritionist has recommended that the pet has at least 30 grams of protein and at least 16 grams of fat per day. Also, the pet should not eat more than 12 ounces of food per day.

Curtis has two types of food available: Food A and Food B. Each ounce of Food A supplies 2 grams of protein and 4 grams of fat. Each ounce of Food B supplies 6 grams of protein and 2 grams of fat.

Curtis wants to vary his pet's diet and stay within these requirements. He needs to know what his options are.

1. Choose variables to represent the amount of each type of food Curtis will include in the pet's daily diet. State clearly what the variables represent.

2. Use your variables to write inequalities to describe the constraints of the problem.

3. Choose one of your constraints. Draw a graph that shows which combinations of Food A and Food B satisfy that constraint. Be sure to label your axes and their scales.

Adapted from *Mathematics with Application* by Lial and Miller. ©1987 by Scott, Foresman and Company. Reprinted by permission of Addison-Wesley Educational Publishers Inc.

Picturing Cookies—Part II

You have graphed each constraint from the unit problem, presented in *How Many of Each Kind?,* on its own set of axes. Each graph shows you a picture of what that constraint means.

Now you will combine the constraints to create one picture of all of them together.

1. Begin with one of the constraints that you worked with before. Using a colored pencil, color the set of points that satisfy this constraint. (Unlike your work on *Picturing Cookies—Part I,* do not color the points that *fail* to satisfy the constraint.)

2. Now choose a second constraint from the problem.
   a. On the same set of axes, using a different color, color the set of points that satisfy this new constraint.
   b. Using your work so far, identify those points that satisfy *both* of the constraints.

3. Continue with the other constraints. Use the same set of axes and a new color for each new constraint.
   a. Color the set of points that satisfy each new constraint.
   b. After graphing each new constraint, identify those points that satisfy all the constraints graphed so far.

4. When you have graphed all the constraints, look at your overall work. Make a single new graph that shows the set of all those points that represent possible combinations of the two types of cookies the Woos can make. In your graph, show all of the lines that come from the constraints. Label each line with its equation.

What's My Inequality?

Graphs of inequalities play an important role in understanding some problem situations. In *Picturing Cookies—Part I,* you started with an algebraic statement—a **linear inequality** from the unit problem—and saw that its graph was a half plane.

Now you will go from graphs back to algebra. In Part I, you are given a graph of a straight line. Your task is to find the related **linear equation.** In Part II, you are given the equation for a line. Your task is to find the inequality corresponding to the half plane on one side of that line.

Part I: Find the Equation

For the straight lines in graphs 1 to 4, write a linear equation whose graph is that straight line. Also describe in words how you found the equation.

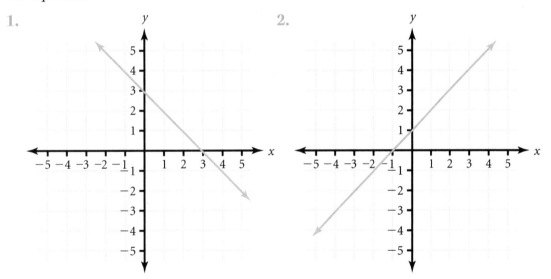

1.

2.

continued

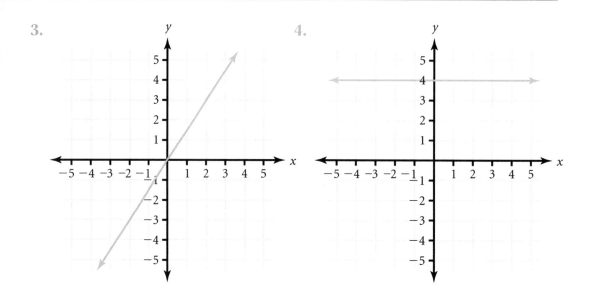

3.

4.

Part II: Find the Inequality

The shaded area in graphs 5 to 8 represents a **half plane.** (Imagine that the shaded area continues indefinitely, including all points on the shaded side of the line.) In each case, you are given an equation for the line that forms the boundary of the half plane. Your task is to find a linear inequality whose graph is the half plane itself.

If the boundary is shown as a dashed line, it is not considered part of the shaded area. If the boundary is shown as a solid line, it is considered part of the shaded area.

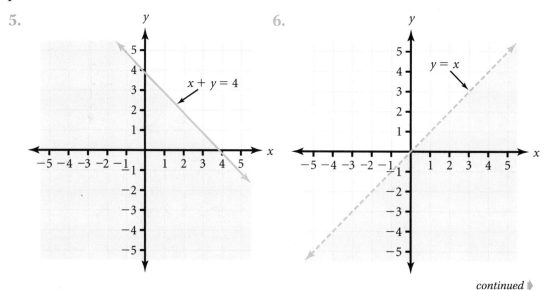

5.

6.

continued ▶

7.

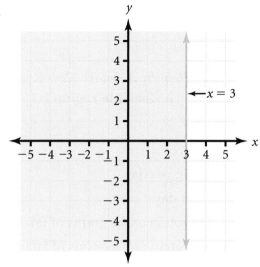

8.

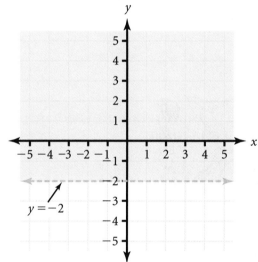

Feasible Diets

You graphed the individual constraints from *Healthy Animals.* Now you will draw the feasible region for that problem. The **feasible region** is the set of points that satisfy the constraints.

Here are the key facts.

- Curtis's pet needs at least 30 grams of protein per day.
- Curtis's pet needs at least 16 grams of fat per day.
- Each ounce of Food A supplies 2 grams of protein and 4 grams of fat.
- Each ounce of Food B supplies 6 grams of protein and 2 grams of fat.
- Curtis's pet should eat no more than 12 ounces of food per day.

Be sure to identify your variables, label the axes, and show the scales on the axes.

Picturing Pictures

Hassan is an artist who specializes in geometric designs. He is getting ready for a street fair next month.

Hassan paints both pastels and watercolors. Each type of picture takes him about the same amount of time to paint. He thinks he has time to make a total of at most 16 pictures.

The materials for each pastel will cost $5. The materials for each watercolor will cost $15. He has $180 to spend on materials. He makes a profit of $40 on each pastel and a profit of $100 on each watercolor.

1. Express Hassan's constraints as inequalities. Use p to represent the number of pastels and w to represent the number of watercolors.

2. Make a graph that shows Hassan's feasible region. The graph should show all the combinations of pastels and watercolors that satisfy his constraints.

3. For at least five points on your graph, find the profit that Hassan would make for that combination.

4. Write an algebraic equation to represent Hassan's profit in terms of p and w.

Using the Feasible Region

As you have seen, the collection of inequalities that describes the unit problem can be represented geometrically as the feasible region. But how do you use this region to solve the problem? How do you determine which point in the region will maximize the Woos' profit?

You will now look at several problems similar to the bakery problem. You will examine how geometry can help you find the maximum or minimum value of a **linear expression** within a feasible region.

Sean Scott uses the graph of a feasible region to determine maximum profit.

Profitable Pictures

Hassan asked his friend Sharma for advice about what combination of pictures to paint. She suggested that he determine a reasonable profit for that month's work. Then he could paint what he needs in order to earn that amount of profit.

Here are the facts about Hassan's paintings.

- Each pastel requires $5 in materials and earns a profit of $40.
- Each watercolor requires $15 in materials and earns a profit of $100.
- Hassan has $180 to spend on materials.
- Hassan can paint at most 16 pictures.

See if you can help Hassan and Sharma. Prepare a written report on the situation. This report should include your work on Questions 1 to 4. Most importantly, it should contain your explanation for Question 5.

1. You have already found the feasible region for the problem. Remember, the feasible region is the set of points that satisfy the constraints. On graph paper, copy this feasible region to use in this activity. Label the axes and show the scales.

continued ▶

2. Suppose Hassan decides that $1,000 would be a reasonable profit.

   a. Find three combinations of watercolors and pastels that would earn Hassan a profit of exactly $1,000.

   b. Mark these three number pairs on your graph.

3. Now suppose Hassan wants to earn only $500 in profit. Find three combinations of watercolors and pastels that will earn a profit of exactly $500. Using a different color, add these points to your graph.

4. Now suppose Hassan wants to earn $600 in profit. Find three combinations of watercolors and pastels that will earn a profit of exactly $600. Using a third color, add these points to your graph.

5. Hassan's mother has convinced him that he should try to earn as much as possible. Hassan wants to figure out the highest profit he can make within his constraints. He also wants to prove to his mother that this profit really is the maximum amount.

   a. Find the maximum possible profit that Hassan can earn and the combination of pictures he needs to paint to earn that profit.

   b. Write an explanation that will convince Hassan's mother that your answer is correct.

Curtis and Hassan Make Choices

1. Curtis goes into the pet store to buy a substantial supply of food for his pet. He sees that Food A costs $2 per pound and Food B costs $3 per pound. Curtis intends to vary his pet's diet from day to day, so he isn't especially concerned about how much of each type of food he buys.

   a. Suppose Curtis has $30 to spend. Find several combinations of the two foods that he might buy. Plot them on an appropriately labeled graph.

   b. Find some combinations that Curtis might buy if he were spending $50. Plot them on the same set of axes.

   c. What do you notice about your answers to parts a and b?

2. Hassan feels there will be a big demand for his work at the street fair. He is considering changing his prices so he earns a profit of $50 on each pastel and $175 on each watercolor.

   a. Based on these new profits, find some combinations of watercolors and pastels so Hassan's total profit would be $700. Plot them on a graph. (The combinations you give here don't have to fit Hassan's usual constraints.)

   b. Repeat part a for a total profit of $1,750, using the same set of axes.

   c. What do you notice about your answers to parts a and b?

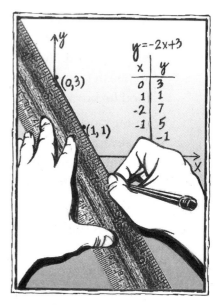

Finding Linear Graphs

Throughout this unit, you are using the graphs of linear equations and inequalities to understand problems. Now you will look at the techniques you use to graph linear equations and perhaps find some shortcuts.

1. One approach to graphing an equation is to make a table of number pairs that fit the equation. Then you graph the number pairs and connect the points with a straight line.

   a. Create a table of at least five number pairs that satisfy the equation $3x + y = 9$.

   b. Plot the number pairs from your table. Connect the points with a straight line.

2. Now graph these equations, looking for shortcuts or special techniques. Pay attention to the methods you use. You will write about your methods in Question 3. (*Note:* Read Question 3 before you do Question 2.)

   a. $y = x + 4$

   b. $x + y = 6$

   c. $2x = 3y$

   d. $2x + 3y = 12$

   e. $5y = 6x - 30$

3. Describe in detail the steps you would take to graph a linear equation. Include any special methods you use that you think might help others. In particular, when you are looking for specific points to plot, how do you decide what numbers to try? If your approach depends on the equation, explain why.

Kick It!

The Free Thinkers football league simply has to do things differently. The folks in this league aren't about to score their games the way everyone else does. So they have thought up a new scoring system.

- Each field goal scores 5 points.
- Each touchdown scores 3 points.

The only way to score points in their league is with field goals, touchdowns, or some combination of field goals and touchdowns.

One of the Free Thinkers, Juliana, has noticed that not every score is possible in their league. For example, a score of 1 point isn't possible, and neither is 2 or 4. But she thinks that beyond a certain number, all scores are possible. In fact, she thinks she knows the highest score that is impossible to make.

1. Figure out what that highest impossible score is for the league. Explain why you are sure that all higher scores are possible.

2. Create another scoring system (using whole numbers). Figure out if any scores are impossible to make in your system. Is there always a highest impossible score? If you think so, explain why. If you think there isn't always a highest impossible score, find a rule for when there is a highest impossible score and when there is not.

3. Consider situations in which there is a highest impossible score. See if you can find any patterns or rules in those situations for figuring out the highest impossible score. You may find patterns that apply in some special cases.

continued

○ *Write-up*

1. *Problem Statement*

2. *Process:* Include a description of any scoring systems you examined other than the one given in the problem.

3. *Conclusions*

   a. State what you decided is the highest impossible score for the Free Thinkers' scoring system. Prove both that this score is impossible and that all higher scores are possible.

   b. Describe any results you found for other systems. Include any general ideas or patterns you discovered that apply to all scoring systems. Prove they apply in general.

4. *Evaluation*

5. *Self-assessment*

Hassan's a Hit!

Hassan's pictures are indeed a big hit, especially the watercolors. Based on his success, he is raising his prices. He will now earn a profit of $50 on each pastel and $175 on each watercolor.

Hassan still has the same constraints: He has only $180 to spend on materials and he can make at most 16 pictures. He had already figured out, using the old prices, how many pictures of each type he should paint to maximize his overall profit.

Hassan now wants to maximize his overall profit with the new prices. Should he paint a different number of pictures of each type? Explain your answer.

You Are What You Eat

The Hernandez twins do not like breakfast. Given a choice, they would rather skip breakfast and concentrate on lunch.

The only things they will eat for breakfast are Fruit-Nuts and Crispies cereals. The twins are allergic to milk, so they eat their cereal dry.

Mr. Hernandez thinks his children should eat breakfast every morning. He also wants their breakfast to be nutritious. Specifically, he would like them each to get at least 5 grams of protein and not more than 50 grams of carbohydrate each morning.

Each ounce of Fruit-Nuts has 2 grams of protein and 15 grams of carbohydrate. Each ounce of Crispies contains 1 gram of protein and 10 grams of carbohydrate.

What's the least amount of cereal each twin can eat while satisfying her father's requirements? Mr. Hernandez wants proof that his criteria are met. The twins want proof that they cannot eat less than this amount of cereal.

Changing What You Eat

In *You Are What You Eat,* you figured out how the Hernandez twins could get enough protein and not too many grams of carbohydrate while eating as little as possible. The answer depended on the ingredients in the cereals and on their father's nutritional requirements.

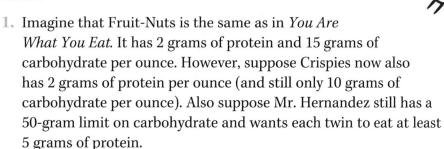

What if the cereals had been a little different? Or if Mr. Hernandez had been stricter about the twins' carbohydrate intake?

1. Imagine that Fruit-Nuts is the same as in *You Are What You Eat.* It has 2 grams of protein and 15 grams of carbohydrate per ounce. However, suppose Crispies now also has 2 grams of protein per ounce (and still only 10 grams of carbohydrate per ounce). Also suppose Mr. Hernandez still has a 50-gram limit on carbohydrate and wants each twin to eat at least 5 grams of protein.

    How much of each cereal should the twins eat? Remember, they want to eat as little as possible.

2. Now suppose Fruit-Nuts has 3 grams of protein and 20 grams of carbohydrate per ounce. Crispies has the same amounts as in Question 1: It has 2 grams of protein and 10 grams of carbohydrate per ounce. Also suppose Mr. Hernandez has now decided the twins can't eat more than 30 grams of carbohydrate each. However, they still need to eat at least 5 grams of protein each.

    What should the twins do now?

Rock 'n' Rap

The Hits on a Shoestring music company is planning for next month. The company makes CDs of both rock and rap music.

Producing a rock CD costs the company an average of $15,000. Producing a rap CD costs the company an average of $12,000. The higher cost for rock comes from needing more instrumentalists for rock CDs. In addition, producing a rock CD takes about 18 hours. Producing a rap CD takes about 25 hours.

The company can afford to spend up to $150,000 on production next month. Also, according to its agreement with the employee union, the company will spend at least 175 hours on production.

Hits on a Shoestring earns $20,000 in profit on each rock CD it produces. It earns $30,000 in profit on each rap CD it produces. But the company recently promised its distributor that it would not release more rap than rock music. The distributor thinks the company is more closely associated with rock music in the public mind.

The company needs to decide how many of each type of CD to make. It is able to make a fractional part of a CD next month and finish it the month after.

1. Graph the feasible region.

2. Explore two different profits.

   a. Find at least three combinations of rock and rap CDs that would give the company a profit of $120,000. Mark these points in one color on your graph. The combinations do not have to be in the feasible region.

   b. Now, using a different color, mark points on your graph that will earn $240,000 in profit.

3. Find out how many CDs of each type the company should make next month to maximize its profit.

4. Explain how you found your answer to Question 3. Also explain why you think your answer gives the maximum profit.

A Rock 'n' Rap Variation

In *Rock 'n' Rap*, you figured out how many rock CDs and how many rap CDs Hits on a Shoestring should produce to maximize its profit.

Suppose the conditions were the same as in that activity except the profits were reversed. In other words, suppose the company makes $30,000 profit on each rock CD and $20,000 profit on each rap CD.

Would this change your advice about how many CDs of each type the company should produce to maximize its profit? If so, how many of each type should the company make? What would Hits on a Shoestring's profit be? Explain how you found your answer.

Getting on Good Terms

Graphing calculators can make finding feasible regions easier. However, to draw the graph of an equation on a graphing calculator, you need to put the equation into "$y =$" form. That is, you need to write the equation so one variable is expressed in terms of the other. For example, you might rewrite the equation $y - 5 = 4x$ as $y = 4x + 5$.

For each equation, express the variable y in terms of the variable x.

1. $y - 2x = 7$

2. $7y = 14x - 21$

3. $5x + 3y = 17$

4. $5(x + 3y) = 2x - 3$

5. $4x - 7y = 2y + 3x$

6. $3y + 7 = 20 - (4x - y)$

Going Out for Lunch

Imagine! You have just started a full-time summer job in an office. It's your first day on the job. The boss has sent you out to buy lunch for the 23 people who work in the office. Everyone wants either one hot dog or one hamburger for lunch.

When you get to Enrico's Express, Enrico asks how many hot dogs and how many hamburgers you want. You were so excited that you forgot to write down how many of each you were supposed to get!

You see that hamburgers cost $1.50 each and hot dogs cost $1.10 each (tax included). You also know that your boss gave you $32.10.

Your Task

Assume $32.10 is the exact amount needed for your purchase.

1. Figure out, in any way you can, how many hot dogs and how many hamburgers you should order.

2. Do you think your answer is the only one possible? Explain why or why not.

Adapted from *Algebra I* by Paul A. Foerster (Reading, MA: Addison-Wesley, 1990).

Points of Intersection

The feasible region for a system of inequalities gives you a picture of the possible solutions. The family of parallel lines helps you see geometrically how to maximize or minimize a linear expression.

The next step to finding the solution is determining the exact coordinates of that maximum or minimum point. That often means finding a common solution for a pair of linear equations. In the main activity, *Get the Point,* you will examine pairs of linear equations. Your task is to develop one or more methods for finding their solutions.

Antonio Lozano is analyzing a feasible region.

Get the Point

In solving problems like the cookie problem, finding the coordinates of the point where two lines intersect is helpful. As you have seen, this is equivalent to finding the solution to a **system** of two linear equations with two variables. You have probably done this already using either guess-and-check or graphing. Your goal in this activity is to develop an algebraic method for finding this solution.

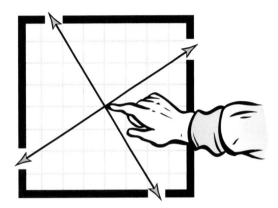

Your written report for this activity should include two things.

- Solutions to Question 1, parts a to e
- The directions your group develops for Question 2

1. For each pair of equations, find the point of intersection of their graphs. Use a method other than graphing or guess-and-check. When you think you have the solution, check it by graphing or by substituting the values into the equations.

    a. $y = 3x$ and $y = 2x + 5$
    b. $y = 4x + 5$ and $y = 3x - 7$
    c. $2x + 3y = 13$ and $y = 4x + 1$
    d. $7x - 3y = 31$ and $y - 5 = 3x$
    e. $4x - 3y = -2$ and $2y + 3 = 3x$

2. As a group, develop general directions for finding the coordinates of the point of intersection of the graphs for two linear equations. Your directions should describe an algebraic method that does not involve guessing or graphing. In developing these instructions, you may want to create some examples like those in Question 1. Your examples can help you come up with ideas or test whether your instructions work.

    Make your instructions easy to follow so someone could use them to "get the point." Your group will also present your ideas to the class.

Only One Variable

In *Get the Point,* your goal is to develop a method for solving systems of two linear equations in two variables. In this activity, you will review the process of solving linear equations in one variable.

1. Solve each linear equation.

    a. $5x + 7 = 24 - 6x$

    b. $6(x - 2) + 5x = 9x - 2(4 - 3x)$

    c. $\dfrac{x + 3}{2} = 29 - 2x$

    d. $\dfrac{3x + 1}{4} = \dfrac{5 - 2x}{6}$

2. Make up a real-world problem that can be represented by a linear equation. Try to create an example in which the variable appears on both sides of the equation.

Set It Up

1. You now have a greater understanding of how to set up and solve pairs of equations in two variables. In this activity you will apply your knowledge.

   Marvelous Marilyn scored 273 points last season for her basketball team. Her points resulted from a combination of 2-point shots and 3-point shots. She made a total of 119 shots. How many shots of each type did Marilyn make?

   Follow these steps to write up this problem.

   - Choose variables and state what they represent.
   - Using your variables, write a pair of equations that represents the problem.
   - Solve the pair of equations graphically.
   - Solve the pair of equations algebraically.
   - Answer the question in the problem.

2. Make up a pair of linear equations whose solution is $x = 3$ and $y = 5$.

A Reflection on Money

1. Uncle Ralph offers you a challenge. He says that if you can tell him how many coins of each type are in his pocket, then you can have the money. He gives you this information.

   • He has 17 coins in his pocket.

   • He has only dimes and quarters.

   • The coins are worth $3.35.

   Can you meet Uncle Ralph's challenge?

   a. Solve the problem graphically.

   b. Solve the problem algebraically.

   c. Describe the advantages and disadvantages of the graphical and algebraic methods.

2. Here are three systems of linear equations. Solve each system algebraically. Use the method you developed in *Get the Point* or any other method you know. Explain your work clearly.

   a. $y = 2x - 3$ and $3x - 4y = 7$

   b. $c + 2f = -6$ and $3c + f = 2$

   c. $2r - k = -1$ and $6r = 5k - 11$

This POW is about solving a puzzle or, rather, about solving a whole set of puzzles. Each puzzle requires two sets of markers, such as two types of coins. Plain and shaded circles are used to represent the markers.

○ *An Example*

One of these puzzles uses three markers of each kind. To begin, the markers are arranged as shown here, with each marker in a square. The plain markers are at the left, and the shaded markers are at the right. There is one empty square in the middle.

The object of the puzzle is to move the markers so the shaded markers end up at the left and the plain markers end up at the right. Here are some rules, of course.

- Plain markers move only to the right. Shaded markers move only to the left.

- A marker can move to an adjacent open square.

- A marker can jump over one marker of the other type into an open square.

- No other types of moves are permitted.

○ *Your Task*

The reason this POW is a *set* of puzzles is that you can vary the number of markers. Your task is to investigate this set of puzzles. Begin with the example just described and answer these questions.

1. Can you solve the puzzle? If so, can you find more than one solution?

continued ◗

2. If you can solve the puzzle, how many moves are you required to make? Is there a minimum number of moves? Can you prove your answer?

Once you have answered these questions for the puzzle with three markers of each type, look at other examples. Consider only cases in which the numbers of each type are equal and exactly one empty square is in the middle. (You can examine other cases in the supplemental activity *Shuttling Variations.*)

Here are some things you can do.

- Figure out if all such puzzles have solutions. If they do, find out how many moves are required.

- Look for a rule that describes the minimum number of moves in terms of the number of markers.

- Prove your results.

○ *Write-up*

1. *Problem Statement*

2. *Process:* Describe how you investigated this set of puzzles. Which cases did you examine? Did you try to develop any general conclusions?

3. *Conclusions and Conjectures:* State the questions you investigated and the conclusions you reached. Include questions that you did not have time to investigate. Some of your conclusions may be about specific cases; others may be more general.

If you can prove any of your conclusions, include the proofs. If any of your conclusions are still tentative, label them as *conjectures.*

4. *Evaluation*

5. *Self-assessment*

More Linear Systems

This activity continues your work with linear equations. Keep alert for new shortcuts and new insights into how to solve one-variable or two-variable equations.

1. Find the solution to each linear equation.

   a. $3(c + 4) - 2c = 16 - 4(c + 5)$

   b. $t + 2(t - 4) = 5(1 - 2t)$

   c. $\dfrac{r + 5}{2} = 12 - 3r$

   d. $7w + 2(3 - 2w) = 4(w + 2) - (w - 6)$

2. Find the values of both variables in each linear system.

   a. $4a - 5b = -4$ and $3a + 6b = 10$

   b. $u - v = 3$ and $2u + 2v = 10$

   c. $2x + 3y = 1$ and $6y = 7 - 4x$

Cookies and the University

You're now ready to solve the unit problem. You will use the feasible region, a family of parallel **profit lines,** and a pair of linear equations to find the cookie combination the Woos should make.

When you're done, you will apply your skills to a completely new problem about college admissions.

Moses Lazo, Anandika Muni, Mandy Mazik, and Vanessa Morales work together to solve the unit problem.

How Many of Each Kind? Revisited

Remember the activity *How Many of Each Kind?* Abby and Bing Woo have a small bakery, and they bake two kinds of cookies—plain cookies and iced cookies. Here is a summary of the key information about the situation.

Summary of the Situation

Facts

- Each dozen plain cookies requires 1 pound of cookie dough.
- Each dozen iced cookies requires 0.7 pound of cookie dough and 0.4 pound of icing.
- Each dozen plain cookies requires 0.1 hour of preparation time.
- Each dozen iced cookies requires 0.15 hour of preparation time.

Constraints

- The Woos have 110 pounds of cookie dough and 32 pounds of icing.
- The Woos have oven space to bake a total of 140 dozen cookies.
- The Woos have 15 hours available to prepare cookies.

Costs, Prices, and Sales

- Plain cookies cost $4.50 a dozen to make. They sell for $6.00 a dozen.
- Iced cookies cost $5.00 a dozen to make. They sell for $7.00 a dozen.
- No matter how many cookies of each kind they make, the Woos will sell them all.

The big question is

How many dozens of each kind of cookie should Abby and Bing make so that their profit is as high as possible?

continued ▶

Your Task

Imagine your group is a business consulting team. The Woos have come to you for help. Of course you want to give them the right answer. But you also want to explain clearly how you know you have found the best possible answer.

You may want to review what you already know. Look back at your notes and your earlier work on this problem. Then write a report for the Woos. Your report should contain these items.

- An answer to the Woos' dilemma. Your answer should include a summary of how much cookie dough, icing, and preparation time they will need. Tell them how many dozens of each kind of cookie they will need to bake. Also tell them how much profit they can expect.

- An explanation that will convince the Woos that your answer will give them the greatest profit.

- Any graphs, charts, equations, or diagrams necessary to support your explanation.

When you write your report, assume the Woos do not know the techniques you have learned in this unit about solving this type of problem.

A Charity Rock

Part I: Solving Systems

Solve each pair of equations.

1. $5x + 2y = 11$ and $x + y = 4$

2. $2p + 5q = 15$ and $6p + 15q = -29$

3. $3a + b = 4$ and $6a + 2b = 8$

Part II: Rocking Pebbles

Rocking Pebbles is a popular rock group. At their concerts, some of the tickets sold are for reserved seats. The rest are for general admission.

The Pebbles put on two concerts last weekend. The band pledged to give their favorite charity an amount equal to half of what was paid for general-admission tickets. After the concerts, the charity called the Pebbles' manager to find out how much money the charity would receive.

The manager looked up the sales. She found that for the first night, 230 reserved-seat tickets and 835 general-admission tickets were sold. For the second night, 250 reserved-seat tickets and 980 general-admission tickets were sold. The manager saw that the total collected for tickets was $23,600 for the first night and $27,100 for the second night. However, she didn't know the prices of the two kinds of tickets. All she knew was that the prices were the same at both concerts.

Figure out what the two ticket prices were. Use that information to tell the manager how much the Rocking Pebbles will donate to the charity. As part of your analysis, set up a pair of linear equations. Solve this system in whatever way you like. You might use algebra, graphs, or guess-and-check.

Back on the Trail

The two situations here are similar to ones you may remember from the Year 1 unit *The Overland Trail*. That unit told the story of pioneers traveling west across the United States in the 1840s.

For each situation, write a pair of equations using two variables. Then solve the equations to answer the question.

Part I: Fair Share on Chores

A family with three boys and two girls needs to split up the chore of watching the animals. Altogether, the animals need to be watched for 10 hours.

The length of each boy's shift is an hour more than the length of each girl's shift. How long is each type of shift? (The family considered this fair in light of other chores the boys and girls had to do.)

Part II: Water Rationing

The Stevens family contains three adults and five children. The Muster family contains two adults and four children. In a typical day on the trail, the Stevenses use about 15 gallons of water for drinking, washing, and so on. The Musters use about 11 gallons of water per day.

Assume each adult uses about the same amount of water. Also assume each child uses about the same amount. How much water does each adult and each child use in a typical day?

Big State U

Big State University needs to decide how many in-state and how many out-of-state students to admit next year. Like all universities, Big State U has limited resources. So, budget concerns play a part in the university's admissions policy.

Here are the constraints on the decision.

- The college president wants this class to contribute at least $2,500,000 to the school after graduation. In the past, Big State U has received an average of $8,000 from each in-state student. It has received about $2,000 from each out-of-state student.

- The faculty wants students with good grade-point averages. Grades of in-state students, on average, are lower than grades of out-of-state students. The faculty is urging the school to admit at least as many out-of-state students as in-state students.

- The housing office can't spend more than $85,000 to cover costs such as meals and utilities for students in dorms during vacation periods. Out-of-state students are more likely to stay on campus during vacations. In-state students cost an average of $100 each for vacation-time expenses. Out-of-state students cost about $200 each.

The college treasurer needs to minimize educational costs. Because students take different courses, teaching an in-state student costs an average of $7,200 a year and teaching an out-of-state student costs about $6,000 a year.

Your job is to recommend how many students from each category Big State U should admit. You need to minimize educational costs, as the treasurer requires. You must do this within the constraints set by the president, the faculty, and the housing office.

Your write-up should include a proof that your solution is the best possible one within the constraints. Show any graphs that seem helpful. Explain your reasoning carefully.

Adapted from *An Introduction to Mathematical Models in the Social and Life Sciences* by Michael Olinick (Reading, MA: Addison-Wesley, 1978).

Inventing Problems

You have now seen several problems you could solve by defining variables and setting up and solving a pair of linear equations. Examples are *Going Out for Lunch, A Charity Rock,* and *Back on the Trail.*

In this activity, you will write your own problem.

1. Make up a problem that you think can be solved with two equations and two unknowns.

2. Solve the problem and write up your solution. As you work on the problem, you may find that you want to change it in some way to improve it.

3. Write your problem (without the solution) on a separate sheet of paper. Put your name on this sheet. In class, you will work on one another's problems.

Creating Problems

Over the course of this unit, you have solved a variety of problems involving linear equations, linear inequalities, and graphs. One way to get more insight into such problems is to create one of your own. In *Inventing Problems,* you created a two-equation/two-unknown problem. In the final days of the unit, you will create a linear programming problem.

Valerie Williams and Bianca Williams work together to analyze a linear programming problem.

Ideas for Linear Programming Problems

In the next activity, *Producing Programming Problems,* your group will create a linear programming problem. You will present your problem to the class. In any **linear programming** problem, you will have variables, constraints, and something to minimize or maximize.

The central problem of this unit (*How Many of Each Kind?*) is a linear programming problem. The variables are the number of dozens of plain cookies and the number of dozens of iced cookies. The constraints are the Woos' available preparation time and oven space and the amounts of cookie dough and icing. The goal is to maximize profit.

Big State U presents another linear programming problem. The variables are the number of in-state students and the number of out-of-state students to be admitted. The constraints involve contributions to the university, grade-point averages, and housing costs. The goal is to minimize educational costs.

Answer Questions 1 to 3 on your own.

1. Reread the activity *You Are What You Eat.*
   a. What do the variables in that activity represent?
   b. What are the constraints?
   c. What needs to be maximized or minimized?

2. Reread the activity *Profitable Pictures.*
   a. What do the variables in that activity represent?
   b. What are the constraints?
   c. What needs to be maximized or minimized?

continued ▶

3. Reread the activity *Rock 'n' Rap*.

   a. What do the variables in that activity represent?

   b. What are the constraints?

   c. What needs to be maximized or minimized?

Now answer Questions 4 and 5 with your group.

4. Create a situation in which you might be interested in maximizing or minimizing something. Describe the situation, and state what you would maximize or minimize.

5. Choose two variables for your situation from Question 4. Give two or three constraints that might apply using those variables.

Producing Programming Problems

Your group will now create a linear programming problem. Here are the key ingredients you need in your problem.

- Two variables
- Something to be maximized or minimized that is a **linear function** of those variables
- Three or four linear constraints

Once you have written your problem, you must solve it.

Then you should put together an interesting five- to ten-minute presentation. This presentation should do three things.

- Explain the problem.
- Provide a solution to the problem.
- Prove that your solution is best.

Beginning Portfolio Selection

The main problem for this unit, presented in *How Many of Each Kind?*, is a linear programming problem. You have seen several such problems, including those in *Profitable Pictures, You Are What You Eat, Rock 'n' Rap,* and *Big State U.*

1. Describe the steps you must go through to solve such a problem.

2. Choose three activities from the unit that helped you to understand particular steps of this process. Explain how each activity helped. You do not need to restrict yourself to the activities listed here.

Selecting these activities and writing your explanations are the first steps toward compiling your portfolio for this unit.

Just for Curiosity's Sake

Part I: Solving Equations

Solve each pair of equations.

1. $3s + t = 13$ and $2s - 4t = 18$
2. $6(a + 2) - b = 31$ and $5a - 2(b - 3) = 23$
3. $z - w = 6$ and $5z + 3w = 10$

Part II: Rocking Pebbles

The Rocking Pebbles just finished another two-night series of concerts. This time, neither show was for charity. Their manager wonders how the producer priced the tickets.

The producer said they sold 200 reserved-seat tickets and 800 general-admission tickets the first night. The total taken in was $20,000. On the second night, they sold 250 reserved-seat tickets and 1000 general-admission tickets. The total taken in that night was $23,000.

Answer the manager's question. Find out the cost of reserved-seat tickets and the cost of general-admission tickets. To do so, write a pair of equations to describe the situation.

Producing Programming Problems Write-up

Your group should now have developed its own linear programming problem. You should also have prepared your group presentation for the class. Now your task is to complete your own write-up for the problem. As with the presentation, your write-up should include three things.

- A statement of the problem
- The solution
- A proof that the solution is the best possible one

Continued Portfolio Selection

In *Beginning Portfolio Selection,* you reviewed one of the major themes of this unit: linear programming problems. Now you will explore another major theme: solving systems of linear equations with two variables.

1. Summarize what you learned about solving such systems. Review your work in *Get the Point* and in later activities.

2. Choose two examples of problem situations you could solve using a system of linear equations. Your examples can be from this unit or from an earlier unit. For each example, explain how the algebraic representation of the problem using linear equations would help you solve the problem.

Cookies Portfolio

Now you will put together your portfolio for *Cookies*. Do these three tasks.

- Write a cover letter that summarizes the unit.
- Choose papers to include from your work in the unit.
- Discuss your personal growth during the unit.

Cover Letter

Review the *Cookies* unit. Describe the unit's central problem and main mathematical ideas. Your description should give an overview of how the key ideas were developed. Also describe how you used them to solve the central problem.

continued ▶

Selecting Papers

Your portfolio for *Cookies* should contain these things.

- *Beginning Portfolio Selection* and *Continued Portfolio Selection*

  Include the activities from the unit that you selected in *Beginning Portfolio Selection* and *Continued Portfolio Selection*, along with your written work on these activities.

- A Problem of the Week

  Select one of the three POWs you completed during this unit: *A Hat of a Different Color, Kick It!,* or *Shuttling Around.*

- *A Reflection on Money*
- *Get the Point*
- *"How Many of Each Kind?" Revisited*
- *"Producing Programming Problems" Write-up*

Personal Growth

Your cover letter should describe how the mathematical ideas were developed in the unit. As part of your portfolio, write about your own personal growth during this unit. You may want to address this question: *How do you think you have improved in your ability to make presentations?*

Also include any other thoughts you wish to share with a reader of your portfolio.

SUPPLEMENTAL ACTIVITIES

Much of this unit concerns the use of graphs to understand real-life problems. The supplemental activities for *Cookies* continue the work with graphs and problem solving. They also follow up on some of the POWs. Here are some examples.

- *Algebra Pictures* and *Find My Region* offer you fun ways to explore graphs of equations and inequalities.

- *Rap Is Hot!* gives you another variation on the situation from *Rock 'n' Rap.*

- *Who Am I?* and *Kick It Harder!* continue the ideas you explored in the POWs *A Hat of a Different Color* and *Kick It!*

Who Am I?

Read the poem below.

At a college class reunion from dear old Big State U,

I met fifteen classmates, counting men and women, too.

More than half were doctors, and the rest all practiced law.

Of the doctors, more were females, and that I clearly saw.

Even more than female doctors were females doing law,

And these statements all would still be true including me, I saw.

If my friend (a noted lawyer) had a wife and kids at home,

Can you draw any conclusions about me from this poem?

State what you can conclude, if anything, about the author of this poem. Show that the line about the friend of the author is needed.

Adapted from *Mathematics: Problem Solving Through Recreational Mathematics* by B. Averbach and O. Chein (New York: Freeman, 1980).

Algebra Pictures

You have been using pictures to help with problems in algebra. Now you will use algebraic inequalities to make artistic pictures—or at least pictures someone might be interested in looking at. But one important difference is that these inequalities are not all linear.

For Questions 1 to 3, make a picture by showing the solution set for the given system of inequalities.

1. $y \le x + 8$
   $y \le 16 - x$
   $y \ge (x - 4)^2$

2. $y \le 2x + 4$
   $y \le 28 - 2x$
   $y \ge 2$
   $y \le 10$

3. $y \le \sqrt{16 - x^2}$
   $y \ge -\sqrt{16 - x^2}$
   $y \le \frac{1}{2}x^2$

4. Create an interesting picture using systems of inequalities. Your picture can be made up of several parts. Each individual part should be the solution set for a system of inequalities.

Find My Region

This activity is a game for two people. The first step is to find a partner.

Setting Up a Feasible Region

Each partner needs to define a feasible region using an inequality.

You should each start with the square region in the first quadrant shown in the graph below left. This region is bounded by the inequalities $x \geq 0$, $x \leq 10$, $y \geq 0$, and $y \leq 10$.

Then each partner needs to choose an inequality to restrict the region further. For example, if you choose the inequality $x + y \leq 15$, your new region will be the shaded area in the graph below right.

Sketch your feasible region on a sheet of graph paper.

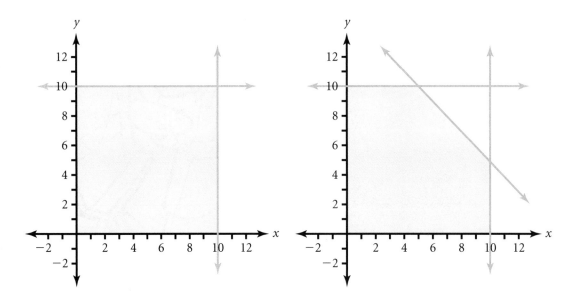

continued ▶

Guessing Each Other's Inequalities

Do not tell your partner what your inequality is or show him or her the region you have created. The goal of the game is to figure out what inequality the other player has used. Sit back to back so that neither of you can see the other's feasible region. You will each need a blank piece of graph paper to keep track of information you gather about your partner's region.

Here are the rules.

- Take turns guessing a point in your partner's feasible region. For example, you might say, "I guess (3, 6)." (Remember you both start with the inequalities $x \geq 0$, $x \leq 10$, $y \geq 0$, and $y \leq 10$. So you should guess only points within the square region that these inequalities define.)

- Each time one of you guesses a point, the other player will say "inside," "outside," or "boundary" to indicate whether the point is inside the region, outside the region, or on a boundary line. For the previous region (9, 6) is a boundary point, (5, 7) is an inside point, and (8, 9) is an outside point.

- When you guess a point, you can choose instead to guess your partner's inequality. For example, you might say, "I guess that $x + y \leq 15$ is your inequality." Your partner must answer truthfully, stating whether your guess matches his or her inequality. If you guess "$x + y \leq 15$" and your partner used "$2x + 2y \leq 30$," your guess is correct. If your guess is incorrect, your partner takes another turn. You do not get a chance to guess a point.

- The winner is the first player to guess the other's inequality correctly.

Advanced Version

You can make this game more challenging by having each player choose two or three inequalities (in addition to the four inequalities defining the square).

Kick It Harder!

In the POW *Kick It!* you probably found that some scoring systems have a highest impossible score and others do not. Now you will explore this issue further.

1. Find a rule that tells you which scoring systems have no highest impossible score.

2. Find a rule that tells you what the highest impossible score is when a scoring system has one.

3. Consider scoring systems that have no highest impossible score. Write a proof of why they have no highest impossible score.

More Cereal Variations

You have explored some problems involving the Hernandez twins and their breakfast habits. Now you will create some variations of your own.

1. Create a variation on the situation in which the twins would choose to eat only Crispies.

2. Create a variation on the situation that does not have a solution.

Rap Is Hot!

The distributor for Hits on a Shoestring has changed her mind about rap. She now believes that rap is more popular in her territory than rock. She tells the company that it can make up to twice as many rap CDs as rock CDs.

The remaining facts are the same as in *Rock 'n' Rap*. Here is a summary of the constraints.

- Producing a rock CD costs an average of $15,000. Producing a rap CD costs about $12,000.

- Producing a rock CD takes about 18 hours. Producing a rap CD takes about 25 hours.

- Hits on a Shoestring must use at least 175 hours of production time.

- Hits on a Shoestring can spend up to $150,000 on production next month.

- Each rock CD makes a profit of $20,000. Each rap CD makes a profit of $30,000.

Find how many CDs of each type Hits on a Shoestring should make next month to maximize its profits. Justify your reasoning. Remember, the company can plan to make a fraction of a CD next month and finish it the month after.

How Low Can You Get?

Anya and Jesse are working together on a problem. They are trying to find which point in a certain feasible region minimizes various linear equations involving x and y. Here are the constraints on the situation.

$$y \geq 3$$

$$y \leq 3x$$

$$y \leq \frac{1}{2}x + 5$$

$$y \leq 23 - 4x$$

Here is the feasible region for this set of constraints.

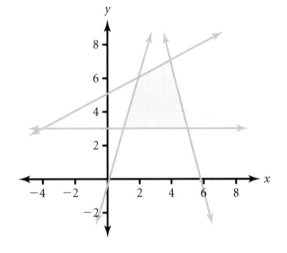

1. Invent two linear expressions in terms of x and y that have different points in the region as their minimum. Explain why the two points really are the minimums for each expression in that region.

2. Invent two different linear expressions that have the same point in the region as their minimum. Explain why the point really is the minimum of each expression in that region.

3. Anya claims that she can pick any point in the region and find a linear expression that has its minimum for the region at that point. Jesse thinks that only certain points can be used as the minimum for a linear expression.

   Who do you think is right? If you agree with Anya, show that any point can be used as a minimum. If you agree with Jesse, explain which points can be used as a minimum and why others cannot.

Shuttling Variations

In the POW *Shuttling Around,* you examined a family of puzzles involving two kinds of markers. In the POW the number of plain markers is equal to the number of shaded markers, and there is exactly one empty square in the middle. Now you will consider variations on this family of puzzles.

What if the number of markers of each type is different? For instance here's a puzzle with three plain markers and four shaded markers.

Also, what if the puzzle has more than one empty square? For instance here's a puzzle with two plain markers, three shaded markers, and two empty squares.

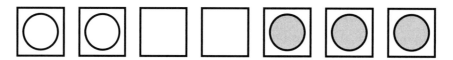

The task is still to move the markers so the shaded markers end up at the left and the plain markers end up at the right. These rules still apply.

• Plain markers move only to the right. Shaded markers move only to the left.

• A marker can move to an adjacent open square.

• A marker can jump over one marker of the other type into an open square.

• No other types of moves are permitted.

Explore these puzzles by doing the following things.

• Determine whether all such puzzles have solutions.

• Look for a rule that describes the minimum number of moves in terms of the number of markers of each type and the number of empty squares.

• Prove your results.

And Then There Were Three

You have done a lot of work in *Cookies* with two linear equations in two unknowns. Now you will extend the ideas and techniques you have learned to systems of three linear equations in three unknowns.

1. Here is a problem to solve with three linear equations in three unknowns.

   I have some dimes, nickels, and quarters. I have 18 coins in all. The total number of dimes and nickels is equal to the number of quarters. The value of my coins is $2.85. How many coins of each kind do I have?

   a. Define your variables carefully.

   b. Write three linear equations that express the facts in the problem.

   c. Solve your system of equations.

2. Make up problems for two other situations that can be solved using three variables and three linear equations.

3. Make up two more systems of three linear equations in three unknowns. Try to solve them. You do not need to create problem situations for these systems.

4. Write down general directions for finding the solution to a system of three linear equations in three unknowns.

An Age-Old Algebra Problem

Bob's, Maria's, and Shoshana's birthdays are on the same day. Bob is two years younger than the sum of Shoshana's and Maria's present ages. In five years, Bob will be twice as old as Maria will be then. Two years ago, Maria was half as old as Shoshana was.

How old is each friend?

1. Set up a system of linear equations for this problem. Be especially careful about defining your variables.

2. Solve your equations using algebra.

Adapted from *Mathematics: Problem Solving Through Recreational Mathematics* by B. Averbach and O. Chein (New York: Freeman, 1980).

Is There Really a Difference?

The Chi-Square Test and the Null Hypothesis

Is There Really a Difference?—The Chi-Square Test and the Null Hypothesis

Data, Data, Data

This unit is about data. More specifically, this unit is about how and when to draw conclusions by comparing sets of data. To begin, you will generate and study some data about yourself and your classmates. You will then investigate ways to represent data in graphical form. You will also learn about the stages in developing and evaluating hypotheses about data.

Hazel Corbi prepares a double-bar graph from data she has collected.

Leigh's parents are concerned about all the time she spends on the phone. They decide to limit her calls to the period from 8:00 p.m. to 9:00 p.m. on school nights. Leigh's friends' parents adopt the same rule.

As long phone conversations seem out of the question, Leigh and her friends need a more efficient way of spreading news. By experimenting, they find that 3 minutes per phone call gives them enough time to tell someone the essentials of most situations. Three minutes includes the time needed to place the call and get the right person on the line.

So here's their plan. The next time Leigh learns something she thinks everyone will want to hear, she will call Michael at exactly 8:00. That call will take 3 minutes. At 8:03, Leigh will call Diane while Michael calls Ana Mai. After they all finish talking, those four will each call someone else and so on.

Your task is to figure out how many of Leigh's friends could hear the news by 9:00 p.m. In analyzing the situation, you should make certain assumptions.

- Making a connection and completing a call always takes 3 minutes.
- No one calls a person who has already been called.
- The caller always reaches the person being called.

○ *Write-up*

1. *Problem Statement*

2. *Process:* Explain how you kept track of your results.

3. *Solution*

4. *Self-assessment*

Samples and Populations

Part I: Picking a Sample

To test a **hypothesis** about a **population,** you need to choose a **sample** that is likely to represent the population accurately. Consider that idea in answering these questions.

1. A music producer wants to find out what high school students think about various kinds of popular music. He conducts a survey of several high school Boy Scout troops to get their opinions.

    a. Do you think Boy Scouts form a good sample of high school students? If not, what might be a more representative sample? Explain your answer.

    b. Give an example of a conclusion the producer might reach based on a survey of Boy Scouts that might not be true about high school students in general.

continued ▶

2. A car manufacturer wants to conduct on-the-street interviews to find out what adults in the United States think of the company's latest TV advertising campaign. The interviewer decides to use a group of people standing at the bus stop near her home as the sample population.

   a. Do you think the people at the bus stop form a good sample of adults in the United States? If not, what might be a more representative sample? Explain your answer.

   b. Give an example of a conclusion the interviewer might reach based on a survey of the people at the bus stop that might not be true about adults in the United States in general.

3. A member of the city council wants to know what people in Middle City think about a proposed new park in the center of town. He picks names in the Middle City phone book at random and calls to get a sample of opinions.

   a. Do you think names chosen at random from the Middle City phone book form a good sample of the city's population? If not, what might be a more representative sample? Explain your answer.

   b. Give an example of a conclusion the councilmember might reach based on a survey of people picked at random from the phone book that might not be true about the city's population in general.

Part II: Literary Sorting

An English teacher conducts a survey asking students about their reading preferences. She lists four book titles and asks students to identify their favorite among those choices.

She surveys sophomores and juniors. Each sophomore puts an *S* next to the title of his or her favorite book on the list. Juniors put a *J* next to their favorite. Here are the results of the survey.

The Grapes of Wrath	J J S J S S J J J S S S J J J J
To Kill a Mockingbird	S J S J S S J S J S S S S J S S
The Bluest Eye	S S S J S S J J S S J J J J J S J J
The Great Gatsby	J J S J S S S J J J S S S S J J J S S S

Make a double-bar graph of this set of data.

Try This Case

Mr. Swenson reads a newspaper ad for a can of mixed nuts from the Fresh Taste company. The ad claims, "Fresh Taste mixed nuts contain peanuts, walnuts, almonds, and cashews in the ratio $4:3:2:1$."

Mr. Swenson is at his local supermarket in no time flat. He buys a can of Fresh Taste mixed nuts and goes home to count the nuts. He finds 300 nuts in the can— 125 peanuts, 98 walnuts, 53 almonds, and 24 cashews. Mr. Swenson wants to sue the Fresh Taste company for false advertising.

1. Imagine you are a lawyer. What arguments would you present if you were representing Mr. Swenson? What arguments would you present if you were representing the Fresh Taste company?

2. Now imagine you are a member of the jury hearing this case. Do you think Mr. Swenson has enough evidence to win his case? Explain why or why not. If you think he doesn't have enough evidence, explain what further evidence you would need before ruling in his favor.

Who Gets A's and Measles?

People often want to know what actions might lead to certain results. One approach to finding out is to study what actions have preceded those results in the past. Read the descriptions of two such studies. Then answer the questions.

Getting A's

Clara wants to find out how to get an A on her next unit assessment. To do so, she has her classmates fill out a questionnaire about their activities before the last unit assessment. Specifically, she asks what they ate and what they did both the night before the assessment and the morning of the assessment. Then she finds out who earned A's on the assessment.

She learns that all the "A" students ate dessert and listened to rap music the night before the assessment. They all drank orange juice the morning of the assessment. None of the other students had done all three things.

Getting Measles

A medical researcher is trying to understand a measles epidemic. She examines the records of 500 patients of various doctors. These records show each person's sex, blood type, cholesterol level, blood pressure, weight, height, and medical history.

The researcher sees that all the patients with measles were overweight and had high blood pressure. She concludes that overweight people with high blood pressure are more likely to contract measles.

The Questions

1. How are the methods used in these two studies the same? How do they differ?

2. What can be concluded from each of the studies? Is either study useful? If so, how and why?

3. How could the research methods be improved?

Quality of Investigation

Part I: Cigarettes and Lung Cancer

The situations in *Who Gets A's and Measles?* illustrate the danger of jumping to conclusions. You should realize that even professional researchers often take unwarranted shortcuts. But don't get the impression that no research study is legitimate or that you can't trust the conclusions of any such study.

You may have heard that many studies link cigarette smoking to lung cancer. Because of this, federal law requires that every package of cigarettes contain a warning about the connection between tobacco smoking and lung cancer. The tobacco industry has argued this is unfair. Its representatives say they can point to studies that do not show such a connection.

What do you think? Are such warnings unfair to tobacco companies? How should you evaluate conflicting studies? Explain your answers.

Part II: Take Your Pick

Choose Option 1 or Option 2.

Option 1

   a. Find an article in a newspaper or magazine that reports on a particular research study or survey.

   b. Summarize the main ideas and conclusions of the study.

   c. Identify information you would like to have about the study that was not included in the article.

Option 2

   a. Use your own environment—such as your school, home, or neighborhood—to gather some data.

   b. Describe a population for which your data could be a sample.

   c. Give an incorrect generalization that someone might make about that population from your data.

Coins and Dice

Coins and dice are good tools for studying probability and data, because you have theoretical models of what to expect.

One of the central problems of this unit involves a coin of questionable fairness. You'll learn about this coin in the activity *Two Different Differences*. You will make your own preliminary decision about whether the coin seems fair. That activity also involves a survey about soft drinks.

In the next few activities, you will use both coins and dice to think about random events. You'll return to the case of the questionable coin—as well as the soft drink survey—near the end of the unit.

These activities also introduce the concept of a **null hypothesis.** This is a kind of "neutral" assumption statisticians often use when investigating data.

Two Different Differences

In each of these two situations, you will investigate what conclusions you can draw from the given data. You will also assess how confident you are about your conclusions.

A Suspicious Coin

Every time there is an extra dessert at Roberto's house, his older brother takes out his special coin. He always lets Roberto flip the coin and always calls out "heads."

Roberto believes that his brother wins more often than he should with a fair coin. One day Roberto finds the coin and flips it 1000 times. He gets 573 heads and 427 tails.

Do you believe the coin is fair? That is, do you think the apparent preference for heads is because the coin is not balanced, or is it just a coincidence?

In your group, try to reach a consensus about this question. Then write a report that includes three items.

- A statement of Roberto's hypothesis
- A statement of your group's conclusion about the coin and an explanation of how you reached that conclusion
- A rating of how confident you are about your conclusion based on a scale of 0 to 10, with 0 meaning "no confidence" and 10 meaning "complete confidence"

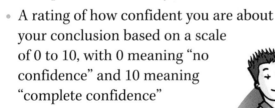

continued

To Market, to Market

A soft drink company has developed a new product. The company's marketing department thinks the new beverage may appeal more to men than to women. The marketing staff need to verify this so they can target their advertising properly.

The marketing staff surveys 150 people at a local supermarket to see how people like the new product. They give each person a sample of both the new soft drink and the old soft drink and ask which drink he or she prefers.

Of the 90 men questioned, 54 prefer the new soft drink and 36 like the old one better. Of the 60 women surveyed, 33 prefer the new soft drink and 27 like the old one better.

Based on these data, do you think the new soft drink will be more popular with men than with women? Try to reach a consensus in your group, and then write a report. The report should include three items.

- A statement of the marketing department's hypothesis
- A statement of your group's conclusion about how men and women view the new beverage and an explanation of how you reached that conclusion
- A rating of how confident you are about your conclusion based on a scale of 0 to 10, with 0 meaning "no confidence" and 10 meaning "complete confidence"

Changing the Difference

Part I: Coins and Soft Drinks

Questions 1 to 5 relate to the activity *Two Different Differences*. Some of these questions have no single right answer.

1. Suppose you flip a fair coin 1000 times. What do you think the probability is of getting exactly 500 heads?

2. Roberto's data in "A Suspicious Coin" might or might not convince you that his brother's coin is unfair. Whether or not you are convinced by his data, give an example of data for 1000 flips that *would* convince you the coin is unfair.

3. Suppose you find a coin and flip it 1000 times. Complete these sentences, replacing each ? with a number.

   a. If the experiment gives between ? and ? heads, then I will accept that the coin is fair.

   b. If the experiment gives between ? and ? heads, then I will suspect the coin is biased in favor of heads.

   c. If the experiment gives more than ? heads, then I will be fairly certain the coin is biased in favor of heads.

continued

4. In "To Market, to Market," the survey contained 90 men and 60 women. For these questions, keep those totals but adjust the number of people within each group who prefer each drink as needed.

   a. Make up data that would convince you men and women like the new soft drink equally.

   b. Make up data that would convince you men like the new soft drink more than women do.

   c. Make up data that would convince you women like the new soft drink more than men do.

5. How does the situation in "A Suspicious Coin" differ from the situation in "To Market, to Market"? How are the situations the same?

Part II: Stick-up Ideas

Make up a question that your class can use for a stick-up graph. This question should compare two populations in some respect and be appropriate for your class.

Questions Without Answers

1. A record company executive is trying to convince a recording artist to go on tour to promote the artist's newest release.

   a. What do you think the executive's hypothesis is about the effect of a tour?

   b. What null hypothesis might the artist propose to avoid the tour?

2. A pharmaceutical company wants to advertise its new product as an anti-acne medicine. The Food and Drug Administration (FDA) is opposed to that plan.

   a. What is the company's hypothesis about its product?

   b. What null hypothesis might the FDA want to test?

3. Some of the owners in a professional basketball league want to raise the basket from 10 feet to 12 feet to make the game more exciting.

   a. What is the hypothesis of the league owners who support the change?

   b. What null hypothesis might the other owners propose?

4. A newspaper marketer is trying to persuade a business owner to place an ad in the Sunday paper.

   a. What hypothesis might the marketer suggest to the business owner?

   b. What null hypothesis might the owner propose instead?

5. Students in a Year 1 IMP class were experimenting to determine what factors influence the period of a pendulum's swing. In their experiment, they changed the weight of their pendulum's bob. They found a different period for their pendulum.

   a. What hypothesis does the experiment suggest?

   b. What null hypothesis should the students consider?

Loaded Dice

Work on this task with a partner from your group. You will be given a pattern for a die. One pair from your group will use the pattern to make a fair die, and one pair will use it to make a loaded (or unfair) die. Before you begin, decide which pair will make which kind of die.

1. Cut out the pattern around the outer solid line.

2. Carefully fold the pattern along all the dashed lines. This task will be easier if you score the lines first with a pen.

3. Figure out how to fold the pattern into a cube with all the dots on the outside. After you have folded the pattern, carefully unfold it.

4. If you are "loading" a die (only one pair of students in your group will do this), carefully tape two regular paper clips to the inside of one of the faces.

5. Fold the die together again, and use transparent tape to hold the cube together.

If time allows, roll your die many times and record the results. Are you getting an even distribution of the numbers 1 through 6?

Fair Dice

1. Roll a regular die 60 times, keeping track of the rolls. Make a frequency bar graph of the data. Then repeat the process and make a separate frequency bar graph of the new data.

   a. Do your two graphs look exactly alike? If not, how do they differ?

   b. Do you think your die is fair? Explain your reasoning.

2. Lucky Lou's Game Shop receives a shipment of loaded and fair dice. The loaded dice are designed to look and feel just like standard dice. Unfortunately, Lou accidentally mixed all the dice together.

   a. Suppose you are trying to figure out which dice are loaded and which are fair by rolling each die numerous times. How many times do you think you will have to roll each die?

   b. What kind of results will you need to determine whether a die is fair or loaded? Explain your reasoning.

Loaded or Not?

You and your partner have been given one of the dice made by your class. It may be a loaded die or it may be a fair die.

1. Answer this question with your partner.

   a. Roll your die 60 times. Make a frequency bar graph of the data.

   b. Compare the frequency bar graphs you and your partner made in *Fair Dice* (two graphs for each of you) with this new graph.

2. Answer this question on your own.

   a. Write about what you observed in comparing the graphs.

   b. Do you think you have a fair die? Explain the basis for your decision.

   c. How confident are you that you correctly identified the die as fair or loaded? Rate your confidence using a scale from 0 to 10, with 0 meaning "no confidence" and 10 meaning "complete confidence." Explain the reason for your rating.

Tying the Knots

Keekerik is an imaginary land where the people have an interesting three-stage ritual for couples who want to get married. Wandalina and Gerik go to the home of Queen Katalana to perform this ritual. Permission for them to marry right away depends on the outcome of the ritual.

○ Stage 1: Loose Ends Top and Bottom

The queen greets the couple. She reaches into a colorful box and pulls out six identical strings. The queen hands the strings to Wandalina, who holds them firmly in her fist. One end of each string is sticking out above Wandalina's fist. The other end of each string is sticking out below her fist.

○ Stage 2: The Tops Are Tied

The queen steps to the side, and Gerik comes forward. He ties two of the ends that are above Wandalina's fist together. Then he ties two other ends above her fist together. Finally, he ties the remaining two ends above her fist together. The six ends below Wandalina's fist still hang untied.

○ Stage 3: The Bottoms Are Tied

Queen Katalana comes forward again. Although she was watching Gerik, she has no idea which string end below Wandalina's fist belongs to which end above. The queen performs the final step. She randomly chooses two of the ends below Wandalina's fist and ties them together, then two more, and finally the last two. Wandalina now has six strings in her fist, with three knots above and three knots below.

continued

Whether Wandalina and Gerik will be allowed to marry right away depends on what happens when Wandalina opens her fist. If the six strings form one large loop, the couple can marry. Otherwise, they will be required to repeat the ritual in six months.

With this in mind, think about these questions.

1. When Wandalina opens her fist, what combinations of different-size loops might there be?

2. What is the probability that the strings will form one large loop? In other words, what are the chances that Wandalina and Gerik will be able to marry right away?

3. What is the probability of each of the other possible combinations?

You may want to do some experiments to get some ideas about these questions. However, your answers for Questions 2 and 3 should involve discussion of the theoretical probability for each result, not just experimental evidence.

○ *Write-up*

1. *Problem Statement*

2. *Process:* Explain how you worked on this problem. Describe any experiments you performed and how you kept track of your results.

3. *Solution:* Give the probability of each possible outcome. Explain how you determined each probability.

4. *Self-assessment*

The Dunking Principle

Mr. Rose, the principal of Bayside High, agrees to participate in a dunking booth at the school fair. Here's how the booth works.

People who purchase tickets will push a button. The button operates a light above the booth. The light is programmed to flash either green or red, using a randomizing mechanism. If the light turns green, Mr. Rose falls into the water. If it turns red, he does not. He is told that the light is set to have a 50% chance of turning green.

As it turns out, Mr. Rose seems to be getting dunked far more often than 50% of the time. In fact, after 20 pushes of the button, he has been dunked 15 times! He is pretty suspicious of what he has gotten himself into.

1. What is Mr. Rose's hypothesis? What is the null hypothesis?

2. Based on the results so far, do you think Mr. Rose is justified in being suspicious?

3. Suppose the fair continues, and Mr. Rose is dunked 46 times out of 60 pushes of the button. Would that convince you the dunking booth is not set the way he was told? What if he is dunked 72 times out of 100 pushes of the button?

4. Mr. Rose's original experience—15 dunks out of 20 pushes— represents 75% dunks. Question 3 asks about other examples involving approximately 75% dunks.

   If you *are not* convinced by either result described in Question 3 that the booth is not set at 50% green, how many occurrences of about 75% dunks would it take to convince you? If you *are* convinced by results like those described in Question 3, what is the smallest 75% dunk result that would convince you?

   Explain your answer in either case.

How Different Is Really Different?

In "A Suspicious Coin" (the problem regarding Roberto's brother's coin described in *Two Different Differences*), you wanted to know if the coin was fair. If you flip a fair coin 1000 times, you would expect to get about 500 heads. But Roberto got 573 heads from his brother's coin. So the question in that problem was this.

*Is this far enough from the **expected number** of heads to conclude the coin is unfair?*

Similar questions are posed here for several different coins.

The Unfairest Coin

Three students—Alberto, Bernard, and Cynthia—each have a coin. They flip their coins different numbers of times and obtain these results.

	Number of heads	Number of tails
Alberto	14	6
Bernard	55	45
Cynthia	460	540

1. Answer this question for each of the three students.
    a. Find the numbers of heads and tails you would expect for that coin if the coin were fair.
    b. Compare your *expected* numbers to the numbers actually *observed*.

2. Suppose you know that exactly one of the three coins is unfair but don't know which coin it is. Based on your analysis in Question 1, which coin would you suspect most? Which would you suspect least? Explain your reasoning.

Whose Is the Unfairest Die?

Alberto, Bernard, and Cynthia have friends who prefer rolling dice to flipping coins. These friends—Xavier, Yarnelle, and Zeppa—were rolling dice recently and keeping track of how many 1s they got. Each friend has one die, and each rolls a different number of times, with these results.

	Number of 1s	Number of other rolls
Xavier	1	29
Yarnelle	23	77
Zeppa	178	822

1. Answer this question for each die.

   a. Find the number of 1s and the number of other rolls you would expect if the die were fair.

   b. Compare these expected numbers to the observed numbers.

2. Suppose you know that one of the three dice is unfair but don't know which die it is. Based on the information from Question 1, which die do you suspect most? Which do you suspect least? Explain your reasoning.

Coin-Flip Graph

In preparation for this activity, you were asked to gather data about the number of heads you got in a 100-coin-flip experiment. You'll now work with the entire class's results.

1. Make a frequency bar graph of the class's coin-flip results. Your graph should show how often each possible number of heads occurred among the fifty 100-flip sets.

2. What percentage of the results have exactly 50 heads?

3. What percentage of the results have 55 or more heads?

4. About 90% of the results are between which two numbers of heads?

A Tool for Measuring Differences

You may recall standard deviation as a tool for studying measurement variation. The next several activities are devoted to developing another tool, called the **chi-square statistic,** for evaluating **sampling fluctuation.** *Chi* is the name of a Greek letter. It is written χ and pronounced like "sky" without the "s."

Like standard deviation, the chi-square statistic involves a complicated computation. But you'll see that learning how to calculate this statistic is just part of the task. The real challenge is learning how to use and how to interpret it.

Annie Tam and Nabil Jiwa use their calculators to find the chi-square statistic.

Normal Distribution and
Standard Deviation

What Is the Normal Distribution?

Many phenomena have a similar distribution pattern—commonly called *bell shaped*—in which a particular result becomes gradually less likely the further it is from the average. The **normal distribution** is a special case of a bell-shaped distribution. It has a precise mathematical definition and frequently can be used as an excellent approximation to real-world situations.

This diagram shows a typical normal curve, in which the height of the curve above a given point on the horizontal axis reflects the likelihood of getting the result marked on that axis. Suppose you select two values on the horizontal axis and draw corresponding vertical lines on the graph. The probability of getting a data result between those two values is the ratio of the area under the curve between those lines compared to the total area under the curve.

What Is Standard Deviation?

The **standard deviation** of a set of data measures how "spread out" the data set is. In other words, it tells you whether the data items bunch together close to the mean or are distributed "all over the place."

The two superimposed graphs here show two normal distributions with the same mean, but the taller graph is less "spread out." The data set represented by the taller graph, therefore, has a smaller standard deviation.

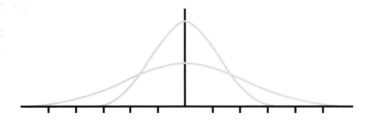

continued

Standard Deviation and the Normal Distribution

One of the reasons standard deviation is so important for normal distributions is that some principles about standard deviation hold true for any normal distribution. Specifically, when a set of data is normally distributed, these statements hold true.

- Approximately 68% of all results are within one standard deviation of the mean.
- Approximately 95% of all results are within two standard deviations of the mean.

These facts can be explained in terms of area, using this diagram.

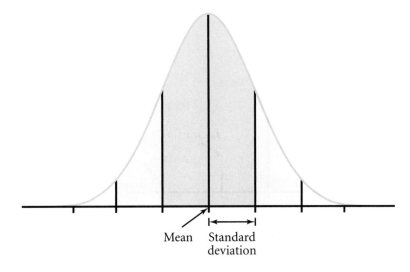

Mean Standard deviation

The Normal Distribution

The darker shaded area in the previous diagram stretches from one standard deviation below the mean to one standard deviation above the mean. This area is approximately 68% of the total area under the curve. The light and dark shaded areas together stretch from two standard deviations below the mean to two standard deviations above the mean. These areas constitute approximately 95% of the total area under the curve. So standard deviation provides a good rule of thumb for deciding whether something is "really different."

Note: To understand where the numbers 68% and 95% come from, you would need a precise definition of normal distribution, which uses concepts from calculus.

continued ▶

Geometric Interpretation of Standard Deviation

Geometrically, the standard deviation for a normal distribution turns out to be the horizontal distance from the mean to the places on the curve where the curve changes from being concave down to concave up. In this next diagram, the center section of the curve, near the mean, is concave down. The two "tails"—the sections farther from the mean—are concave up.

The two places where the curve changes its concavity are marked by vertical lines. They are exactly one standard deviation from the mean, measured horizontally.

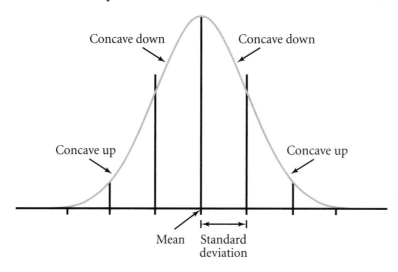

Visualizing the standard deviation

Calculation of Standard Deviation

Here are the steps for calculating standard deviation.

1. Find the mean.

2. Find the difference between each data item and the mean.

3. Square each of the differences.

4. Find the mean (average) of these squared differences.

5. Take the square root of this average.

continued ▶

To calculate standard deviation, organizing the data set into a table like the one that follows can be very helpful. This table is based on a data set of five items: 5, 8, 10, 14, and 18. The mean of this data set is 11. The mean of a set of data is often represented by the symbol $\bar{x}$, which is read "x bar."

The computation of the mean is shown below the table. Step 4 of the computation of the standard deviation, shown below the table to the right, is broken down into two substeps: (1) adding the squares of the differences and (2) dividing by the number of data items. The symbol generally used for standard deviation is the lowercase form of the Greek letter *sigma*, written σ.

x	$x - \bar{x}$	$(x - \bar{x})^2$
5	-6	36
8	-3	9
10	-1	1
14	3	9
18	7	49

sum of the data items = 55   sum of the squared differences = 104

number of data items = 5   mean of the squared differences = 20.8

$\bar{x}$ (mean of the data items) = 11 σ (standard deviation) = $\sqrt{20.8} \approx 4.6$

Suppose you represent the mean as $\bar{x}$, use n for the number of data items, and represent the data items as x_1, x_2, and so on. The standard deviation can then be defined by the equation

$$\sigma = \sqrt{\frac{\sum_{i=1}^{n}(x_i - x)^2}{n}}$$

Bacterial Culture

Scientists at a research lab are trying to determine whether a new product called Infect-Away is better at killing bacteria than Bact-Out, the current market leader.

The researchers have tested Bact-Out extensively, applying it to a bacterial culture of a fixed size and measuring the number of bacteria left after a fixed time. They have seen that the number of bacteria left appears to be normally distributed. The average number of bacteria left using Bact-Out is 1000, with a standard deviation of 50.

1. Sketch this normal distribution. Show the mean and the approximate location of the first and second standard deviation borders.

2. Replace each ? in this sentence.

    If the bacterial test is performed over and over, one would expect about 95% of the results to show between ? and ? bacteria left after Bact-Out is applied.

To evaluate Infect-Away, the researchers use the same procedure as for Bact-Out.

3. a. What is their null hypothesis?

    b. What might their hypothesis be?

4. In testing Infect-Away, suppose the researchers find the culture has 825 bacteria left. What should they conclude and why?

Decisions with Deviation

1. The Typhoon Twister water ride has a minimum height requirement of 5 feet. The average height of children who visit this amusement park is 4 feet 10 inches, with a standard deviation of 2 inches.

   Assume the distribution of children's heights is approximately normal. Approximately what percentage of the children who come to the park qualify to ride the Typhoon Twister? Explain.

2. Lob and Smash is a company that manufactures tennis balls. The company wants to get a contract to provide the Big Bucks Tennis League with balls for its tournaments.

   The tennis league wants tight control over the bounce of the tennis balls it uses. The league has a standard "bounce test." It wants balls that bounce approximately 3 feet (when dropped from a certain height), varying no more than 2 inches from this amount.

   The bounce of Lob and Smash tennis balls is approximately normally distributed. Using the league's test, these tennis balls are shown to have a mean bounce height of 2 feet 11 inches, with a standard deviation of 0.5 inch. Should the Big Bucks Tennis League use Lob and Smash tennis balls? Why or why not?

3. The time it takes Sam to get out of bed, get dressed, eat, and catch the bus varies from morning to morning. Measuring from the moment she wakes up until she reaches the bus stop, her routine takes an average of 15 minutes, with a standard deviation of 5 minutes. Assume the duration of her routine is normally distributed.

   Sam likes to catch the 7:00 a.m. bus so she can visit with her friends before school. What time should she wake up to catch this bus about 84% of the time?

The Spoon or the Coin?

Roberto isn't sure whether his brother's coin is fair or not, but he certainly doesn't like the results he has obtained in the past. He tells his brother he will no longer have the fate of the extra dessert determined by that coin.

Roberto's brother smiles and gives Roberto a new option. They will toss ten spoons in the air and let them fall on the floor.

- If more than five spoons land "bowl down," Roberto will get the extra dessert.

- If more than five spoons land "bowl up," Roberto's brother will get the extra dessert.

- If the spoons are evenly split—five up and five down—the brothers will toss the spoons again.

1. Put yourself in Roberto's place. Conduct a set of experiments that will help you decide whether to use this method for deciding who gets the extra dessert.

2. What would you conclude from your experiments? How confident are you of your conclusion?

Measuring Weirdness

The results from Roberto's brother's coin made Roberto suspect that it isn't a fair coin.

Standard deviation is one tool for measuring how unusual a result is, especially in a situation involving a normal (or almost normal) distribution. But you don't always know the standard deviation for a situation, and many distributions are not normal. So other "measures of weirdness" are needed besides standard deviation.

In this activity, you will look at ways to measure "weirdness" for coin-flip results.

Here are the data about the three coins from the activity *How Different Is Really Different?*

	Number of heads	Number of tails
Alberto	14	6
Bernard	55	45
Cynthia	460	540

In that activity, you compared each student's "observed numbers" to the "expected numbers" and decided which coin seems to be the most "suspicious." In other words, you investigated which coin result would be the "weirdest" for a fair coin.

Alberto and Bernard have each come up with a way to measure the weirdness of the coin flips.

1. Alberto's method is to find the numeric difference between the observed number of heads and the expected number of heads. Because Alberto doesn't care whether the weirdness shows up as too many or too few heads, he subtracts the smaller number from the larger number. For example, with Cynthia's coin, he subtracts 460 from 500 to get 40.

   a. Using Alberto's method, rank the three coins from most to least weird.

continued

b. Explain, using examples, why Alberto's method is inadequate for deciding which of two results is more unusual. In other words, create two possible sets of coin-flip results for which the one that seems obviously "weirder" actually has a smaller numeric difference. Use different sample sizes for the two sets of results.

2. Bernard's method is to use the percentage difference. That is, he finds the difference between the observed percentage of heads and the expected percentage of heads. For example, with Cynthia's coin, the percentage difference is 4%, because 46% of her flips are heads whereas the expected value is 50% heads.

   a. Using Bernard's method, rank the three coins from most to least weird.

   b. Explain, using examples, why Bernard's method is also inadequate for deciding which of two results is more unusual. In this case, create two possible sets of coin-flip results for which the one that seems obviously weirder actually has a smaller percentage difference. Again, use different sample sizes.

3. Try to come up with a third method that Cynthia could use for measuring weirdness that you think is better than either Alberto's or Bernard's method.

Drug Dragnet: Fair or Foul?

People in many occupations, such as police officers and air traffic controllers, are subject to random drug testing. Such testing is done on the grounds that these employees' work affects the safety of the general public.

Drug testing is controversial, however. One objection concerns the potential unfair consequences of the fact that the tests are imperfect. For instance, if a test incorrectly shows someone to be a drug user (a "false positive"), that person could lose his or her job. In this activity, you will explore some of the mathematical issues in drug testing.

Assume a certain test for drug use is 98% accurate. This means 98% of people who use the given drug within some specified time period will test positive and 98% of the people who did not use the drug in that time period will test negative. Also assume that only 5% of people on the job (1 in every 20) engage in drug use.

1. If someone tests positive, how likely is it that he or she has actually engaged in drug use within the given time period? To answer this question, consider a large population such as 10,000 people. Figure out how many people in that population use drugs and how many users and nonusers test positive.

2. Do you think such a test should be used? Explain.

How Does χ^2 Work?

You have seen that the χ^2, or chi-square, statistic is computed using the expression

$$\frac{(\text{observed} - \text{expected})^2}{\text{expected}}$$

This expression is computed for each observed number. The χ^2 statistic is defined to be the sum of these results. For example, in coin-flip problems, there are two observed numbers: the number of heads and the number of tails.

Now that you know how to do the chi-square calculation, you need to understand how it reflects the actual situation. What does this magic number tell you?

Your group will investigate what different χ^2 numbers mean. You will do this by making up coin-flip results and calculating the χ^2 statistic for them.

1. Choose a sample size and make up some coin-flip results that might occur for that sample size. Calculate the χ^2 statistic for each result. Then describe what happens to the value of the χ^2 statistic as the results get "weirder."

You saw in *Measuring Weirdness* that Alberto's and Bernard's methods—numeric difference and percentage difference—don't necessarily tell you which of two results is weirder if the sample sizes are different. Questions 2 and 3 focus on whether the χ^2 statistic is sensitive to changes in sample size.

2. Consider these two cases.

   - You flip a coin 1000 times and get 510 heads.
   - You flip a coin 30 times and get 25 heads.

   In both cases, you have 10 heads more than expected. So Alberto's method would rate them as equally weird.

   a. Find the χ^2 statistic for each case.

   b. Do the χ^2 statistics reflect your intuition about which case is weirder?

continued ▸

3. Consider these two cases.

   - You flip a coin 12 times and get 3 heads.
   - You flip a coin 60 times and get 15 heads.

   In both cases, you have 25% fewer heads than expected. So Bernard's method would rate them as equally weird.

   a. Find the χ^2 statistic for each case.

   b. Do the χ^2 statistics reflect your intuition about which case is weirder?

4. What numeric value for the χ^2 statistic would lead you to believe that a coin-flip result is really unusual? Explain.

The Same χ^2

Getting 40 heads and 10 tails out of 50 flips of a fair coin would be pretty weird, right? Well, it turns out that the χ^2 statistic for that result is 18. The expected number for both heads and tails is 25, and

$$\chi^2 = \frac{(40 - 25)^2}{25} + \frac{(10 - 25)^2}{25} = 18$$

This activity is about other coin-flip results that have a χ^2 statistic of 18.

1. Suppose you flip a fair coin 100 times.

   a. Based on your intuition, decide how many heads and how many tails out of 100 flips would seem just as weird as 40 heads and 10 tails out of 50 flips.

   b. Calculate the χ^2 statistic for that 100-flip result.

   c. By trial and error, find the number of heads and tails out of 100 flips that actually comes closest to giving a χ^2 statistic of 18.

2. Repeat Question 1 for 1000 flips.

3. What generalizations can you make based on your answers to Questions 1 and 2?

Measuring Weirdness with χ^2

In *Measuring Weirdness,* you
looked at Alberto's and Bernard's
methods for deciding which of
the coins from *How Different Is
Really Different?* is the weirdest.
Now you will look at what the χ^2
statistic has to say about those
coins. Then you'll examine the
case of "A Suspicious Coin"
(Roberto's brother's coin in
Two Different Differences).

For your convenience, here are the results of the three coins.

	Number of heads	Number of tails
Alberto	14	6
Bernard	55	45
Cynthia	460	540

1. Do parts a and b for each of the three coins.

   a. Find the number of heads and the number of tails you would
   expect if the coin were fair.

   b. Calculate the χ^2 statistic. Remember, there are two observed
   numbers for coin-flip problems: the number of heads and the
   number of tails.

2. Suppose you know that one of the three coins is unfair. Based
   on your analysis in Question 1, which coin do you suspect most?
   Which do you suspect least? Explain your answers using χ^2 statistics.

3. Calculate the χ^2 statistic for Roberto's brother's coin, which gave
   573 heads out of 1000 flips. Do you think his coin is unfair? Explain.

χ^2 for Dice

In *Measuring Weirdness with χ^2*, you applied the χ^2 statistic to some earlier coin problems. Now you will apply it to the dice data from *Whose Is the Unfairest Die?* You will see how well χ^2 fits your intuition about the weirdness of the dice results.

To refresh your memory, three friends—Xavier, Yarnelle, and Zeppa—were rolling dice and keeping track of how many 1s they got. They each had a different die. They rolled their dice different numbers of times, with these results.

	Number of 1s	Number of other rolls
Xavier	1	29
Yarnelle	23	77
Zeppa	178	822

1. Answer this question for each of the three dice.

   a. Find the number of 1s and the number of other rolls you would expect if the die were fair.

   b. Calculate the χ^2 statistic.

2. Suppose you know that one of the three dice is unfair. Based on the information from Question 1, which die do you suspect most? Which do you suspect least? Explain your reasoning.

3. How do the χ^2 results compare to your intuition?

Does Age Matter?

Clementina waits on tables at an ice cream shop after school. She earns the minimum wage, so a large part of her income is from tips.

Most of her customers are high school students. Clementina thinks that adults are more likely to be good tippers than high school students. She reasons that adults have more money to spend and a better idea of how much to tip.

Clementina's mother has waited on tables for many years at various restaurants. All of her customers are adults, and she has kept track of how many tip well. She says that in her experience, 70% of adult customers tip well.

Clementina has just been offered a job waiting on tables in a coffee shop with nearly all adult customers. She is trying to decide whether to take the new position. She's had fun working with high school students, but she thinks she might make more money at the new job.

She decides to check out her theory about tipping. She keeps careful track of her tips for an afternoon at the ice cream shop. Out of 52 high school customers, she receives 30 good tips and 22 poor tips.

1. What is Clementina's hypothesis? What is the null hypothesis?

2. Suppose high school customers have the same tipping habits as the adult customers in Clementina's mother's experience. What would the expected number of good tips for Clementina's survey be?

3. Calculate the χ^2 statistic for Clementina's data.

4. Do you think Clementina would earn more in tips if she changed jobs? Why?

5. If you were Clementina, would you change jobs? Explain your reasoning.

Different Flips

You will now begin gathering data on the probability of getting certain values of the χ^2 statistic if the null hypothesis is actually true. Knowing such probabilities will allow you to use the χ^2 statistic to decide whether to accept or reject a null hypothesis.

Please do not "fake" your data set. Because you will use this data set to find the approximate probability of getting certain χ^2 statistics, having authentic information on the variety of results that actually occur is important.

1. Flip a coin 50 times. Record the number of heads and tails you get. Write down how many heads and tails you would expect if the coin were fair. Find the χ^2 statistic for your data. Round your answer to the nearest hundredth.

2. Do a similar experiment, this time with 40 flips. Calculate a second χ^2 statistic, rounding to the nearest hundredth.

3. Do a third experiment, this time with 60 flips. Calculate a third χ^2 statistic, rounding to the nearest hundredth.

4. Are your χ^2 statistics the same? Why or why not?

Graphing the Difference

1. Make a frequency bar graph of the class data from *Different Flips*.

2. What percentage of the class data have a χ^2 statistic greater than 1?

3. Based on your graph, estimate the probability of getting a χ^2 statistic greater than 3, assuming the null hypothesis is true.

4. Estimate the probability of getting a χ^2 statistic less than 4, assuming the null hypothesis is true.

Assigning Probabilities

1. In *Measuring Weirdness with χ^2*, you probably found that the χ^2 statistic for Bernard's coin was 1 (exactly). In *Graphing the Difference*, you made a frequency bar graph of χ^2 statistics from your class data.

   a. How many data items did you use altogether to make your frequency bar graph?

   b. How many of those data items are at least as large as Bernard's χ^2 statistic?

   c. Use your answers to estimate the probability of getting a χ^2 statistic at least as large as Bernard's, assuming the null hypothesis is true.

2. Write general instructions for how to use your χ^2 frequency bar graph to estimate probabilities of coin-flip results.

3. Compare the χ^2 statistic to standard deviation. How are the two measures the same? How are they different?

Random but Fair

Do this activity with a partner.

In this activity, you will use your calculator's random number generator to simulate the null hypothesis from *Does Age Matter?* As you may recall, Clementina's mother, an experienced server, says that 70% of adults are good tippers. The null hypothesis is that high school students fit the same model. So what you want is something like a coin that comes up heads 70% of the time.

A random number generator can act just like that coin.

- If the first digit of the random number—that is, the number in the tenths place—is 0, 1, or 2, the customer was not a good tipper.
- If the first digit is 3, 4, 5, 6, 7, 8, or 9, the customer was a good tipper.

In the long run, this method should give good tips 70% of the time.

1. In the first experiment, you will generate 40 random numbers. This is like having a sample of 40 people. Have one partner use the calculator and report the random numbers, while the other records the results.

   With each random number, the person with the calculator will say "bad" if the digit in the tenths place is 0, 1, or 2. He or she will say "good" if the digit is 3, 4, 5, 6, 7, 8, or 9. The partner will tally the good and bad tippers and announce when 40 results are tallied.

2. Calculate the χ^2 statistic for your result. Use the 70% figure to compute your expected numbers.

3. Repeat Questions 1 and 2 for a sample of 50 people. Then repeat them for a sample of 60 people.

4. Switch roles of reporter and recorder, and repeat Questions 1 to 3. When you are finished, you and your partner will have calculated six χ^2 statistics.

A χ^2 Probability Table

Value of the χ^2 statistic	Probability of getting a χ^2 statistic this large or larger when the null hypothesis is true
0.0	1.0000
0.2	.6547
0.4	.5271
0.6	.4386
0.8	.3711
1.0	.3173
1.2	.2733
1.4	.2367
1.6	.2059
1.8	.1797
2.0	.1573
2.2	.1380
2.4	.1213
2.6	.1069
2.8	.0943
3.0	.0832
3.2	.0736
3.4	.0652
3.6	.0578
3.8	.0513
4.0	.0455
4.2	.0404
4.4	.0359
4.6	.0320
4.8	.0285

continued ▶

5.0	.0254
5.2	.0226
5.4	.0201
5.6	.0180
5.8	.0160
6.0	.0143
6.2	.0128
6.4	.0114
6.6	.0102
6.8	.0091
7.0	.0082
7.2	.0073
7.4	.0065
7.6	.0058
7.8	.0052
8.0	less than .005

A Collection of Coins

1. Al and Betty have two coins. One coin is fair and the other is not. The friends are trying to figure out which coin is which. Betty flips the first coin for a while and gets 40% heads.

   Now Al takes a turn. He flips the other coin a few times. Then he stops, thinks, and does some arithmetic. Even though Al has flipped heads only 25% of the time, he claims he is quite sure his coin is the fair coin—and he's right! How does he know?

2. Find the χ^2 statistic for each of the three samples of coin flips given in parts a, b, and c. Then use the information in *A χ^2 Probability Table* to find the probability that you would get a result that far or farther from what you would expect if the coin were fair.

   a. 25 heads and 35 tails

   b. 220 heads and 240 tails

   c. 995 heads and 1025 tails

3. Make up a sample with the same difference between the number of heads and the number of tails as in Question 2a (namely, a difference of 10) but with a smaller χ^2 statistic than for Question 2a. You do not have to calculate the χ^2 statistic if you can clearly explain how you know it is smaller.

4. Make up a sample with the same number of total flips as in Question 2c (namely, 2020 total flips) but with a larger χ^2 statistic than for Question 2c. You do not have to calculate the χ^2 statistic if you can clearly explain how you know it is larger.

A Difference Investigation

In this POW, you will make a hypothesis about two populations that you think are different in some respect. You will collect data from samples of each population. Then you will examine the sample data and use your analysis to evaluate your hypothesis.

Work on this POW with a partner. You and your partner will create a poster of your results and make a presentation to the class. Each of you will also prepare a write-up of the POW.

○ *Stages of the POW*

This POW has six stages.

1. Tell your teacher who your partner is.

2. You and your partner hand in
   - a precise statement of your hypothesis
   - a statement of your null hypothesis
   - a method for collecting sample data from each of your two populations (If your method of collecting data involves a questionnaire, include a copy of the questionnaire.)

3. You and your partner collect data. Take care that your sample is not biased, that participants do not influence one another, and that you get accurate information.

4. You and your partner analyze the results and prepare a poster and a presentation.

5. Prepare your own write-up of the POW. The parts of your write-up are described in the Write-up section.

6. You and your partner make your presentation using the poster you both created.

continued

○ *Presentation Poster*

Your presentation will include a poster that summarizes your analysis. The poster will contain these items.

- The question you investigated
- The data you collected
- Your table of observed and expected values
- The calculation of the χ^2 statistic
- The probability associated with that χ^2 statistic
- Your conclusions

○ *Write-up*

1. *Statement of Hypothesis*
   - Explain why you chose your hypothesis. That is, explain why you thought the hypothesis was true before you collected data.
   - State the null hypothesis.

2. *Data Collection:* Describe how you collected your data. In particular, answer these three questions.
   - How did you try to guarantee that your sample would be representative of the entire population concerned in your hypothesis? That is, how did you try to avoid **bias?**
   - What did you do to keep participants from influencing one another? That is, how did you keep participants *independent?*
   - Do you think participants were willing to give you accurate information? If not, why not?

3. *Analysis:* On the basis of your sample data, do you think your hypothesis is true, false, or still not proven? Explain your reasoning.

   The important part of this section of your report is your reasoning. Base your reasoning on an analysis of the collected data. In your analysis, include a double-bar graph and a graph showing where your χ^2 statistic fits on the χ^2 distribution curve.

4. *Evaluation:* What did you learn from this POW? Discuss both what you learned about the process of analyzing data and what you learned about the topic you studied.

Late in the Day

The manager of a manufacturing company is concerned about the number of on-the-job accidents that happen. She suspects that more accidents happen at the end of a shift when workers are tired.

If this is true, she will consider changing work schedules in some way to improve safety. However, schedule changes are costly and aren't popular with most workers.

At this company, each employee works an 8-hour shift. The manager studies the company's recent accident records. She finds that of the last 168 accidents, 57 occurred in the last 2 hours of a shift.

1. The manager's hypothesis is that the accident rate is higher late in a shift. What is the null hypothesis?

2. How many accidents out of 168 would you expect to happen in the last 2 hours of a shift if accidents are equally distributed throughout shifts?

3. Suppose you are a statistician hired by the manager. Conduct a χ^2 test with her data. On the basis of your findings, advise her on whether to change schedules. Explain your reasoning carefully because she has never studied statistics.

4. What else should you consider in analyzing this situation?

Comparing Populations

Coins and dice may be fun to play with, but many of us find it more interesting to study people. In these activities, you will see how to use the chi-square statistic to compare two populations. You will also return to the two central problems from *Two Different Differences*.

In addition, you will work on the POW *A Difference Investigation*. For this major project, you will gather data and apply many of the ideas from this unit.

Johnny Tran and Jonathan Wong work together to analyze their chi-square data.

What Would You Expect?

Consider this question.

Who is more likely to be left-handed—a man or a woman?

In this activity, you will look at what you would expect from a survey *if there were no difference* between men and women in this respect.

In other words, your null hypothesis is that men and women are the same with respect to "handedness." You will imagine that you are conducting a survey to help you decide whether to reject this null hypothesis.

Throughout this activity, complete the tables as if the null hypothesis were true.

1. Assume you have a group of 100 women and 100 men. In this group of 200, 150 people are right-handed and 50 are left-handed.

   Make a table like the one shown here. Fill in the empty cells with the number of people you'd expect to find in each of the four groups if the null hypothesis were true—that is, if gender makes no difference to handedness.

	Women	Men	Total
Right-handed			150
Left-handed			50
Total	100	100	200

2. Now assume you have a different group of people. This one contains 120 women and 80 men. In this group of 200, 150 people are right-handed and 50 are left-handed.

continued ▶

Again, make a table like this one. Fill in the number of people you'd expect to find in each of the four groups, assuming the null hypothesis is true.

	Women	Men	Total
Right-handed			150
Left-handed			50
Total	120	80	200

3. Assume you have a third group of people. This one consists of 25 women and 75 men. In this group of 100, 65 people are right-handed and 35 are left-handed.

Make a table like this one. Fill in the number of people you'd expect to find in each of the four groups, assuming the null hypothesis is true.

	Women	Men	Total
Right-handed			65
Left-handed			35
Total	25	75	100

4. Now invent your own group of people. Decide how many men, how many women, how many right-handed people, and how many left-handed people are in your group. Because you are considering one overall group of people, the total number of men and women must equal the total number of left-handed and right-handed people.

Make a table like the others, and fill in the number of people you'd expect to find in each of the four groups.

As in Questions 1 to 3, assume the null hypothesis is true—that is, that gender makes no difference to handedness.

Who's Absent?

An administrator at Bayside High forms the hypothesis that *there is a difference* between tenth graders and eleventh graders with respect to absences from school.

For each question, make a table like the one given. Fill it out based on the assumption that the null hypothesis is true. In other words, throughout this activity, assume there is no difference between tenth graders and eleventh graders at Bayside High with respect to the likelihood of being absent.

1. Assume the school has 200 tenth graders and 200 eleventh graders. Out of those 400 students, 40 are absent. Show how many students you would expect to find in each of the four groups.

	Absent	Not absent	Total
Tenth graders			200
Eleventh graders			200
Total	40	360	400

continued ▶

2. Suppose instead the school has 200 tenth graders and 300 eleventh graders. This time, 60 students are absent. Show how many students you would expect to find in each of the four groups.

	Absent	Not absent	Total
Tenth graders			200
Eleventh graders			300
Total	60	440	500

3. Choose your own values for the numbers of tenth graders and eleventh graders in the school. Also decide on the fraction of the total group that is absent.

Make a new table. Fill in the numbers of tenth graders and eleventh graders you would expect to be absent and the numbers of tenth graders and eleventh graders you would expect not to be absent. Base your figures on the assumption that grade level makes no difference in absence rate.

Big and Strong

The doctors in this activity have made hypotheses about the effects of the care they give their patients.

1. Dr. Eileen Bertram is an obstetrician. She thinks she gives better prenatal care to her patients than most doctors do. As a result, she believes the babies she delivers are less likely to be underweight at birth.

   Last year she delivered 75 babies. Other obstetricians in her building delivered 280 babies. Out of the total of 355 babies, 43 were underweight.

   a. State an appropriate null hypothesis for Dr. Bertram's theory.

   b. Copy the table shown here. Fill in the totals, including the number of babies altogether.

   c. Fill in the rest of the table by finding the expected numbers for this situation. In other words, figure out what the table would look like if the results fit the null hypothesis exactly.

	Underweight babies	Not underweight babies	Total
Dr. Bertram			
Other doctors			
Total			

continued ▶

2. Dr. Oliver Pine is a pediatrician. He has kept track of the children he and his colleagues have cared for. He thinks his patients are more likely to turn out to be good athletes than his colleagues' patients.

Altogether, he has kept track of 121 of his patients and 348 of his colleagues' patients. Out of the total of 469 children, 217 became good athletes.

a. State an appropriate null hypothesis for Dr. Pine's theory.

b. Copy the table shown here. Fill in the totals, including the number of children altogether.

c. Fill in the rest of the table by finding the expected numbers for this situation.

	Good athletes	Not good athletes	Total
Dr. Pine			
Other doctors			
Total			

Delivering Results

In *Big and Strong*, you considered two situations. Each situation compared the patients of one doctor with the patients of his or her colleagues. In each case, you wrote a null hypothesis. Then you completed tables to show what would be expected if the null hypothesis were true.

Now you will make up results to complete the tables in a way that would support a claim different from the null hypothesis. The totals for each table are entered for you.

1. Recall Dr. Bertram's belief that the babies she delivers are less likely to be underweight than the babies other doctors deliver. Make up numbers for the missing entries in the table that would support her theory. Be sure your numbers fit the totals.

	Underweight babies	Not underweight babies	Total
Dr. Bertram			75
Other doctors			280
Total	43	312	355

continued ▶

2. Recall Dr. Pine's belief that the children he cares for are more likely to turn into good athletes than the children other doctors care for. Make up numbers for the missing entries in the table that would support his claim. Be sure your numbers fit the totals.

	Good athletes	Not good athletes	Total
Dr. Pine			121
Other doctors			348
Total	217	252	469

3. You've seen how the χ^2 statistic is defined in situations in which a population is compared to a theoretical model. In the situations from *Big and Strong,* the comparison is between two actual populations. For example, you compared Dr. Bertram's patients with the patients of other doctors.

How do you think the χ^2 statistic might be calculated in a situation like this? For example, how might you calculate the χ^2 statistic for the data you made up in Question 1 of this activity?

Paper or Plastic?

Earlier in this unit, you used the χ^2 statistic to find out if a given population fit a certain theoretical model. In recent activities, you have compared two populations with each other. The central question has been whether the two populations are really different. Now you're ready to use the χ^2 statistic to help answer such questions.

The Situation

A checker at a supermarket thinks people who buy frozen dinners are more likely to prefer plastic bags to paper bags than people who don't buy frozen dinners.

To check his theory (and to make his job more interesting), he keeps track of people going through his checkout stand. He now has a record of 2000 people: 500 who bought at least one frozen dinner and 1500 who bought no frozen dinners.

Of those who bought frozen dinners, 260 requested plastic and 240 requested paper. Of those who did not buy frozen dinners, 740 requested plastic and 760 requested paper.

Do you think these data support the checker's hypothesis that there is a difference in the two groups of people with respect to preference for paper or plastic bags?

The Report

Prepare a report about this situation.

- State what two populations you are comparing.
- State your hypothesis and the null hypothesis.
- Calculate the expected numbers. In other words, decide what numbers you'd expect if the null hypothesis were true.
- Calculate the χ^2 statistic.
- Use the χ^2 probability table to find the probability of getting a χ^2 statistic that large or larger.
- Based on the probability, decide whether to reject the null hypothesis. That is, decide whether you believe the two populations are really different. Explain your reasoning.

Is It Really Worth It?

Feline leukemia is a deadly disease among cats. However, protecting them with a vaccine may be possible. Dr. Lee, a veterinarian, is considering recommending a certain vaccine to his clients for their cats, based on what he has read about its use in another region.

Sometime after the vaccination of cats began in that region, an outbreak of feline leukemia occurred. A report reviewed data on 400 cats in the area of the outbreak. Of those cats, 100 had been vaccinated and 300 had not. The number in each group that actually got sick is shown in the table.

	Got sick	Did not get sick	Total
Vaccinated	10	90	100
Not vaccinated	50	250	300
Total	60	340	400

Dr. Lee wants to know whether he should recommend that his clients bring in their cats to get this vaccine.

Notifying all his clients will be expensive and scheduling the visits will be difficult. Also, the vaccine will cost each client $40, which many would rather not have to spend. And many cat owners would prefer not to subject their pets to the trauma of vaccination.

Yet this disease is the most common cause of death among cats in Dr. Lee's practice. What do you think he should do and why?

Be sure to state a null hypothesis and carry out a χ^2 test. Remember to use the row and column totals—not the observed numbers in the table cells—when you find the expected numbers. Also explain why you think the χ^2 statistic is or is not useful in this case.

Two Different Differences Revisited

In *Two Different Differences,* you looked at two situations—one involving a possibly biased coin and one comparing men's and women's beverage preferences. At that time, you were only able to evaluate the situations intuitively. Now you will use the χ^2 statistic to make more informed judgments about those situations.

Write a report for each situation. Your report should include six items.

- A statement of the populations involved
- A hypothesis
- A null hypothesis
- An explanation of how you found the χ^2 statistic
- A statement of your conclusion and how you used the χ^2 statistic to reach that conclusion
- A rating of how confident you are of your conclusion based on a scale of 0 to 10, with 0 meaning "no confidence" and 10 meaning "complete confidence"

A Suspicious Coin

Roberto's brother has a special coin that he uses whenever there is an extra dessert. Roberto's brother always calls "heads."

Roberto suspects the coin comes up heads more often than it should. One day, Roberto finds the coin and flips it 1000 times. He gets 573 heads and 427 tails.

The question to examine is

Is Roberto's brother's coin biased in favor of heads?

continued ▶

To Market, to Market

The marketing department of a soft drink company suspects its new beverage might appeal more to men than to women.

To test this hypothesis, the department surveys 150 people at a local supermarket to see how they like it compared to the company's old product.

Of the 90 men questioned, 54 prefer the new soft drink and 36 like the old one better. Of the 60 women surveyed, 33 prefer the new soft drink and 27 like the old one better.

The question to examine is

Will the new soft drink be more successful among men than among women?

Reaction Time

Buck Adams has always heard that people should not drink alcoholic beverages and then drive. He isn't sure, however, whether everyone knows how significantly alcohol impairs driving ability.

As a service to his community, Buck conducts an experiment, using a driving simulator to test people's reflexes. He has some participants use the simulator while sober and others while intoxicated beyond the legal limit.

Buck records the number of participants in each category who "crashed" and the number who did not. Here are his results.

	Crashed	Didn't crash
Sober	52	125
Intoxicated	66	88

1. State Buck's hypothesis and the null hypothesis for this experiment.

2. Use the χ^2 statistic to find the probability that Buck's samples of sober drivers and intoxicated drivers would have this great a difference in driving performance if alcohol did not affect driving ability.

3. There is an abundance of warnings against driving while intoxicated and severe penalties for doing so. Why do you think some people still drive while intoxicated?

Bad Research

Read the following "journal entry" from a "researcher." List all the mistakes you can find in the researcher's work and explanation.

Personal Journal Entry 710

I was thumbing through my associate's files today. I came across some files of students he taught at the Junior Academy. According to the records, 5 of the 50 boys in the class failed. Amazingly, no records showed any girls failing!

Assuming he taught equal numbers of boys and girls, I proceeded to calculate the ex-square statistic to see if there was really a difference in the students' failure rates.

I first made this table.

	Boys	Girls	Total
Passed	Expected: 47.5 Observed: 45	Expected: 47.5 Observed: 50	95
Failed	Expected: 2.5 Observed: 5	Expected: 2.5 Observed: 0	5
Total	50	50	100

Then I proceeded to do my calculations.

$$\frac{(45-47.5)^2}{45} + \frac{(50-47.5)^2}{50} + \frac{(2.5-5)^2}{5} + \frac{(2.5-0)^2}{0} = 0.14 + 0.13 + 1.25 + 0$$

$$= 1.52$$

When I looked up 1.52 on the probability table, I found that this ex-square statistic happens about 22% of the time.

I conclude that 22% of boys fail more than girls. More importantly, this proves that women are smarter than men.

On Tour with χ^2

These situations come from *Questions Without Answers.*

1. A record company executive and a recording artist are discussing whether the artist should go on a tour to promote the artist's newest release. The issue is whether there is a difference in record sales when an artist goes on tour.

   To address this issue, they collect some data. They find that of 50 acts that toured, 20 made a profit on their release's sales. Of the 120 acts that did not tour, 30 made a profit.

   a. What are the hypothesis and the null hypothesis in this situation?

   b. Set up a table, and calculate the χ^2 statistic. Find the probability associated with that χ^2 statistic.

   c. Do you think these results are convincing enough to justify having the artist tour?

   d. What are some problems with this study? How would you change the study?

2. The owners in a professional basketball league are talking about whether to raise the basket from 10 feet to 12 feet to make the game more exciting. What the owners really want to know is whether there is a difference in attendance when the basket is set at 12 feet.

   The owners have the basket raised to 12 feet for some exhibition games. They leave the basket at 10 feet for other exhibition games. They then compare attendance at each game with that of the previous year.

continued ▶

Of the 40 games played with a 12-foot basket, 35 show an increase in attendance. Of the 100 games played with a 10-foot basket, 70 show an increase in attendance.

a. What are the hypothesis and the null hypothesis in this situation?

b. Set up a table, and calculate the χ^2 statistic. Find the probability associated with that χ^2 statistic.

c. Do you think these results are convincing enough to justify changing the height of the basket at regular games?

d. What are some problems with this study? How would you change the study?

POW Studies

The end of this unit is devoted to the POW *A Difference Investigation*. You will complete your report and listen to your classmates' presentations.

In the last two activities, you will compile your unit portfolio.

Ceanna Shira begins her POW presentation by explaining the hypothesis she and her partner investigated.

Wrapping It Up

Your task is to complete your POW report. Meet with your partner so you will be prepared to make your presentation.

Beginning Portfolio Selection

You've had an opportunity to use the χ^2 statistic in a number of situations. Now review your work for this unit.

1. Choose one activity for which you feel using the χ^2 statistic is helpful for making a decision.

2. Choose another activity for which you feel the χ^2 statistic is not useful (or at least less useful) for making a decision.

3. Describe the kind of situation for which you feel the χ^2 statistic is most useful for making decisions.

Is There Really a Difference? Portfolio

The time has come for you to assemble your portfolio for the unit. This task has three parts.

- Writing a cover letter summarizing the unit
- Choosing papers to include from your work in this unit
- Discussing the role of the POW *A Difference Investigation* in the unit

Cover Letter

Review *Is There Really a Difference?* and describe the main mathematical ideas in this unit. Your description should give an overview of how the key ideas were developed. In compiling your portfolio, you will select activities that you think were important in developing those key ideas. Your cover letter should include an explanation of why you selected particular items.

Selecting Papers

Your portfolio for *Is There Really a Difference?* should contain these items.

- *"Two Different Differences" Revisited*
- *Beginning Portfolio Selection*
- *POW 9: A Difference Investigation*
- Another Problem of the Week

  Select one of the first two POWs you completed during this unit (*A Timely Phone Tree* or *Tying the Knots*).

- Other high-quality work

  Select one or two other pieces of work that represent your best effort. These can be any work from the unit, such as POWs, homework, classwork, or presentations.

continued

The Role of the POW Project

Discuss how doing *A Difference Investigation* contributed to your understanding of the unit. You may want to think about two aspects in particular.

- The project's role in helping you understand mathematical ideas
- The project's role in helping you appreciate the usefulness or significance of the mathematical ideas

You might also comment on your experience working with a partner on an extended project.

SUPPLEMENTAL ACTIVITIES

Learning how to compute and interpret the chi-square statistic is the heart of this unit. Many of the supplemental activities are designed to deepen your understanding of this tool and expand your ability to use it. Here are some examples.

- In *Explaining χ^2 Behavior*, your goal is to explain how the formula for the χ^2 statistic takes into account variations due to sample size.

- In *Completing the Table*, you will learn about an important technique for "filling in the gaps" in tables like *A χ^2 Probability Table*.

- *Degrees of Freedom* and *A 2 Is a 2 Is a 2, or Is It?* examine how to apply the χ^2 statistic to situations that are more complex than those you've studied before.

Two Calls Each

In the POW *A Timely Phone Tree,* Leigh and her friends were able to keep making phone calls right up until 9:00 p.m. But suppose the parents of all these students decided that no one should make more than two calls a night. How would that affect their phone tree?

In other words, Leigh would still call Michael from 8:00 until 8:03, and then Leigh and Michael would call Diane and Ana Mai from 8:03 to 8:06. But at 8:06, Leigh would be finished. Only the other three friends would be able to make a call at 8:06. Once Michael made his second call, he'd also be finished.

Under these new rules, how many of Leigh's friends could get her news by 9:00 p.m.?

Make the same assumptions as in *A Timely Phone Tree.*

- Making a connection and completing a call always takes 3 minutes.
- No one calls a person who has already been called.
- The caller always reaches the person being called.

Incomplete Reports

Newspapers, television shows, magazines, journals, and other media often include surveys and statistics in their reports. In these reports, you might get a little information about the statistics or a lot of information. Such questions as "How many people did you survey?" get answered in some reports and not in others.

Your task is to investigate what types of information are given in various surveys. Find surveys in which practically all the information is revealed, surveys in which so little information is given that you might question the results, and surveys in between the two extremes.

Some places you can find surveys are

- medical journals and Web sites
- technical documents
- science and health magazines and Web sites
- news periodicals

Comment in detail on at least one of the surveys. Describe the strengths and weaknesses of the survey. Also discuss what information an article should present about a survey for the results to influence your opinion.

If possible, attach copies of the surveys.

Smokers and Emphysema

Emphysema is a serious lung disease that makes it difficult to breathe.

According to one source, 0.86% of people who smoke develop emphysema, while 0.24% of nonsmokers develop it. (*Note:* Both of these figures are less than 1%.)

Based on the assumption that 15% of adults are smokers, answer these questions.

1. What percentage of the population will probably develop emphysema?

2. What percentage of the population will be smokers with emphysema?

3. Given that someone has emphysema, what is the probability that the person is a smoker?

Explaining χ^2 Behavior

You know that for a fair coin, getting 60% heads out of 100 flips is much more unusual than getting 60% heads out of only 10 flips. In other words, getting a result of 60 heads and 40 tails is "weirder" than getting a result of 6 heads and 4 tails.

One of the strengths of the χ^2 statistic is that it reflects this difference.

1. To illustrate this aspect of the χ^2 statistic, do these two computations.

   a. Calculate the χ^2 statistic for an experiment in which you get 60 heads in 100 flips.

   b. Calculate the χ^2 statistic for an experiment in which you get 6 heads in 10 flips.

2. Now examine the formula by which the χ^2 statistic is computed. Show how this formula leads to getting such different χ^2 statistics in Question 1, even though both experiments result in 60% heads.

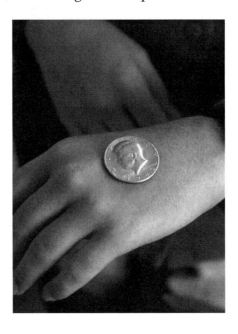

Completing the Table

The χ^2 probability table tells you the probability of getting a χ^2 statistic of a certain value or greater when the null hypothesis is true. However, the table only lists some specific values for the χ^2 statistic.

For example, the probability associated with $\chi^2 = 0.6$ is .4386. The probability associated with $\chi^2 = 0.8$ is .3711. In other words, about 44% of χ^2 statistics are 0.6 or greater, whereas about 37% are 0.8 or greater. But what can you say about a χ^2 statistic of 0.7? What is the probability of getting a χ^2 statistic of 0.7 or greater?

An Intuitive Idea

Intuition might lead you to reason something like this.

> A χ^2 statistic of 0.7 is exactly halfway between $\chi^2 = 0.6$ and $\chi^2 = 0.8$. So the associated probability should be exactly halfway between .4386 and .3711.

1. Test this proposed method.
   a. Look in the χ^2 probability table (from *A χ^2 Probability Table*) to find the probability associated with $\chi^2 = 2.2$ and the probability associated with $\chi^2 = 2.6$.
   b. Find the number exactly halfway between those two probabilities.
   c. Compare your answer for part b with the probability in the table for $\chi^2 = 2.4$. Are they the same? Are they close?

What Actually Happens

As you discovered, the probability in the table for $\chi^2 = 2.4$ is not exactly halfway between the probability for $\chi^2 = 2.2$ and $\chi^2 = 2.6$. It's actually a little closer to the probability for $\chi^2 = 2.6$ than to the probability for $\chi^2 = 2.2$. But it's pretty close to halfway.

continued ▶

The method you just used is a general approximation technique called *linear interpolation*. This technique is especially useful for functions that can't be computed in any simple way, such as the function associating probabilities with a given χ^2 statistic.

2. Here's an illustration of this technique with a much simpler function. Consider the *squaring function*, that is, the function whose equation is $y = x^2$. This function can be represented by the notation $f(x) = x^2$.

   a. Find y when $x = 7$. In other words, find $f(7)$.

   b. Find y when $x = 9$. In other words, find $f(9)$.

   c. Using the technique of linear interpolation, you would expect $f(8)$ to be halfway between $f(7)$ and $f(9)$. Check this out. That is, find the number halfway between $f(7)$ and $f(9)$, and compare it to $f(8)$.

One strength of the linear interpolation technique is that it isn't limited to "halfway" points.

3. Use linear interpolation to find the probability associated with a χ^2 statistic of 0.85. (*Hint:* The value 0.85 is one-fourth of the way from 0.8 to 1.0. Find the probability that lies one-fourth of the way from .3711 to .3173. Keep in mind the probability decreases as the χ^2 statistic increases.)

4. Develop a general formula for using linear interpolation based on this situation.

   Assume you have a function g and you know the values of $g(a)$ and $g(b)$. Also assume that c is a number between a and b.

   What would you use as an estimate for the value of $g(c)$? First find a formula in terms of a, b, and c that tells "what fraction of the way c is from a to b." You might want to look at numeric examples for ideas. For example, why is 0.85 "one-fourth of the way" from 0.8 to 1.0?

Bigger Tables

In some situations comparing populations, the data can be described using a 2-by-2 table. More complex situations require more complicated tables.

In this activity, you will explore what such tables might look like and how you might use them.

1. A reporter wants to compare high school students in different grades in terms of their participation rate in sports. He gathers sample data from several high schools in a major city.

   Before working out the numbers, the reporter set up this table to reflect the totals for the groups he surveyed.

	Ninth graders	Tenth graders	Eleventh graders	Twelfth graders	Total
Plays a sport					400
Does not play a sport					505
Total	250	230	220	205	905

   Assume there is no difference from class to class in terms of the sports participation rate. What would be the expected numbers for the individual cells of this table?

2. When the reporter tallies his information, the table turns out like this.

	Ninth graders	Tenth graders	Eleventh graders	Twelfth graders	Total
Plays a sport	120	110	92	78	400
Does not play a sport	130	120	128	127	505
Total	250	230	220	205	905

continued ▶

Based on his table, do you think there is a significant difference among the classes in sports participation? If so, how do you think the classes are different? In either case, explain your thinking.

3. The reporter's question is whether different classes have different sports participation rates. This question requires a 2-by-4 table.

   a. Develop five questions that would each require a table larger than 2 by 2. Make your questions about situations that are as different as you can make them.

   b. Create data for two of your situations.

   ○ For one situation, create data demonstrating that the underlying populations are significantly different.

   ○ In the other situation, create data for which any differences could easily be attributed to **sampling fluctuation.**

TV Time

A researcher wants to know whether there is a difference in the television-viewing habits of married and single people.

She works with a sample of 200 people, of whom 100 are married and 100 are single. Of the married people, she classifies 35 as light television watchers and 65 as heavy watchers. Of the single people, she classifies 15 as light watchers and 85 as heavy watchers.

Do you think the researcher should conclude there really is a difference between married and single people with respect to television-viewing habits? Explain your answer using the χ^2 statistic.

Degrees of Freedom

You have seen that a 2-by-2 table can be used to compare two populations in terms of whether they possess a given characteristic.

One Degree of Freedom

A 2-by-2 table is said to have "one degree of freedom." If one of the cell numbers is known, all the rest can be calculated from that number (assuming you know the row and column totals).

For instance, look at the table in Question 1 concerning school attendance. You are given just one of the four cell values—namely, that 15 tenth graders were absent.

1. Copy this table. Then, based on this one cell value (and the row and column totals), fill in the rest of the table. Explain your answers.

	Tenth graders	Twelfth graders	Total
Attended			400
Absent	15		35
Total	230	205	435

More Degrees of Freedom

More information is needed to complete tables larger than 2 by 2. The minimum number of cell values needed to complete a table tells you the degrees of freedom for the situation.

Now you will investigate what types of tables represent various degrees of freedom. The goal is to find a general rule that will tell you the degrees of freedom of an *m*-by-*n* table and explain why the rule works.

continued

2. What is the least number of cell values you need to know to complete the table below? That is, how many degrees of freedom are in the situation? Explain your answer.

	Ninth graders	Tenth graders	Twelfth graders	Total
Works after school				630
Does not work after school				550
Total	430	390	360	1180

3. What is the least number of cell values you need to know to complete the table below? That is, how many degrees of freedom are in the situation? Explain your answer.

	Ninth graders	Tenth graders	Eleventh graders	Twelfth graders	Total
Plays a sport					400
Does not play a sport					505
Total	250	230	220	205	905

4. Create more tables, make up row and column totals, and see how many degrees of freedom each has. Include examples with more than two rows. Be sure to check that the sum of the column totals equals the sum of the row totals.

5. Find a rule to calculate the degrees of freedom of an m-by-n table. Explain why your rule works.

A 2 Is a 2 Is a 2, or Is It?

Generalizing the Computation

The χ^2 statistic is used not only for situations with one degree of freedom but also for tables larger than 2 by 2.

The calculation of the χ^2 statistic for more degrees of freedom is essentially the same as that for one degree of freedom.

- Find the value of the expression $\dfrac{(\text{observed} - \text{expected})^2}{\text{expected}}$ for each cell.
- Add these values for all the cells.

The only difference in the computation for a larger table is that there are more cells. That means there are more χ^2 expressions to sum.

Using the χ^2 Statistic

To use the χ^2 statistic, you need to know which χ^2 statistics would be rare and which would be common if the null hypothesis were true.

In *Is There Really a Difference?* you investigated the distribution of χ^2 statistics for the case of one degree of freedom. The results are summarized in your χ^2 probability table. For example, you can look up $\chi^2 = 2.0$ and see that the probability of getting a χ^2 statistic of 2.0 or greater (when the null hypothesis is true) is approximately .1573.

In this activity, you will compare the distribution of χ^2 statistics for situations with more degrees of freedom to the case of one degree of freedom.

continued ▶

A More General χ^2 Probability Table

The table at the end of this activity gives χ^2 probabilities for various degrees of freedom. The column headings represent a fixed list of probabilities. Each row is a χ^2 probability table for a particular number of degrees of freedom.

For example, in the row for one degree of freedom is an entry of 1.64. The heading for this column is ".20" (that is, 20%). This means that for a situation with one degree of freedom, when the null hypothesis is true, 20% of all χ^2 statistics will be 1.64 or greater. This is similar to the entry in your χ^2 probability table, which gives an associated probability of .2059 for a χ^2 statistic of 1.6.

Similarly, in the row for four degrees of freedom is an entry of 1.06 under the column heading ".90." This means that for a situation with four degrees of freedom, when the null hypothesis is true, 90% of experiments will give a χ^2 statistic greater than or equal to 1.06.

1. Using this table (and a lot of estimation), give a general description of how the probability of getting a χ^2 statistic of 2.0 or more changes as the number of degrees of freedom increases.

2. a. Choose a probability "cutoff" for deciding that differences are significant. That is, decide how rare a χ^2 statistic needs to be for you to reject the null hypothesis.

   b. Based on your decision in part a, determine, for each degree of freedom in the table, how large the χ^2 statistic needs to be for you to reject the null hypothesis. Explain how you determined that value.

continued ▶

3. Consider the data in this table.

	Ninth graders	Tenth graders	Eleventh graders	Twelfth graders	Total
Plays a sport	120	110	92	78	400
Does not play a sport	130	120	128	127	505
Total	250	230	220	205	905

a. Calculate the χ^2 statistic for this data set.

b. Based on your result, do you think the different grades participate in sports at the same rate? Explain your answer in terms of the χ^2 statistic and the table.

χ^2 **Probability Table for Different Degrees of Freedom**									
Degree of freedom	Probability								
	.99	.95	.90	.50	.20	.10	.05	.01	.001
1	0.00	0.00	0.02	0.45	1.64	2.71	3.84	6.63	10.83
2	0.02	0.10	0.21	1.39	3.22	4.61	5.99	9.21	13.82
3	0.11	0.35	0.58	2.37	4.64	6.25	7.81	11.34	16.27
4	0.30	0.71	1.06	3.36	5.99	7.78	9.49	13.28	18.47
5	0.55	1.15	1.61	4.35	7.29	9.24	11.07	15.09	20.52
6	0.87	1.64	2.20	5.35	8.56	10.64	12.59	16.81	22.46
7	1.24	2.17	2.83	6.35	9.80	12.02	14.07	18.48	24.32
8	1.65	2.73	3.49	7.34	11.03	13.36	15.51	20.09	26.12
9	2.09	3.33	4.17	8.34	12.24	14.68	16.92	21.67	27.88
10	2.56	3.94	4.87	9.34	13.44	15.99	18.31	23.21	29.59

Fireworks

Quadratic Functions, Graphs, and Equations

Fireworks—Quadratic Functions, Graphs, and Equations

A Quadratic Rocket

You have worked extensively with linear functions of the form $y = ax + b$ and their graphs. The next level of complexity is the **quadratic** function, $y = ax^2 + bx + c$.

This unit launches your exploration of the world of quadratic functions. In the activities, you will focus on the graphs of quadratic functions. The graph of any quadratic function is a special shape called a parabola.

The central problem of this unit involves a rocket launched for a fireworks display. The rocket's path can be modeled by a quadratic function.

Aminta Badian Kouyate writes a quadratic function on the board.

Victory Celebration

The Bayside High varsity soccer team has just won the championship. To celebrate, the school will be putting on a fireworks display. The team is helping with the planning.

The fireworks will travel on a rocket launched from the top of a tower. The top of the tower is 160 feet off the ground. The mechanism will launch the rocket so it will rise initially at 92 feet per second.

The team members want the fireworks to explode when the rocket is at the top of its trajectory. To set the timing mechanism, they need to know how long the rocket will take to reach the top. They also want to tell spectators the best place to stand to see the display, so they need to know how high the rocket will go.

Finally, the team wants to know how long the rocket will be in the air, from the moment it leaves the tower until it comes back down and hits the ground. The entire area is level, so when the rocket lands, it will be at the same height as the base of the tower.

A Rocket Formula

Antonio is on the soccer team. His older sister Hana has studied some physics and mathematics that would be helpful to the team.

She says there is a function $h(t)$ that will give the rocket's height off the ground in terms of the time t since the launch. If t is in seconds and $h(t)$ is in feet, then

$$h(t) = 160 + 92t - 16t^2$$

You can probably see where the numbers 160 and 92 come from. The 16 in the last term comes from the force of gravity.

continued

Your Task

Your task is to help the soccer team find the answers to its questions.

1. Sketch the situation.

2. Write a clear statement of the questions the soccer team wants to answer.

3. Describe how you might use Hana's formula to help answer the team's questions.

4. Using whatever methods you choose, try to get some answers (or partial answers) to the team's questions.

Growth of Rat Populations

Two rats, one male and one female, scampered on board a ship anchored at a local dock. The ship set sail across the ocean. It reached a deserted island in late December. The rats abandoned the ship to make their home on the island.

Given these ideal conditions, you can estimate the number of offspring produced from this pair of rats in one year. To do so, make these four assumptions.

- The original female gives birth to six young on January 1. She produces another litter of six rats every 40 days thereafter as long as she lives.

- Each female rat born on the island will produce her first litter of six young 120 days after her birth. She will produce a new litter of six rats every 40 days thereafter.

- Every litter has three males and three females.

- The rats have no natural enemies on the island and plenty of food. This means that no rats die in the first year.

How many rats will live on the island by the following January 1, including the original pair?

continued

○ *Write-up*

1. *Problem Statement*

2. *Process:* Include a discussion of how you organized your information. Also describe any approaches you tried that didn't work.

3. *Solution:* Explain how you kept track of the results and how you arrived at your answer.

4. *Self-assessment:* Include a statement of how confident you are about the correctness of your answer.

A Corral Variation

Dairyman Johnson raises cattle. A long, straight fence runs along the border between his property and the property of his neighbor, rancher Gonzales.

Dairyman Johnson likes rectangles, and he also values efficiency. He realizes that the fence can serve as part of a pen for his cattle.

His plan is to use part of this border fence as one side of a rectangular pen. He wants to build the other three sides using 500 feet of fencing he has purchased. For example, he might build the pen as shown in the diagram.

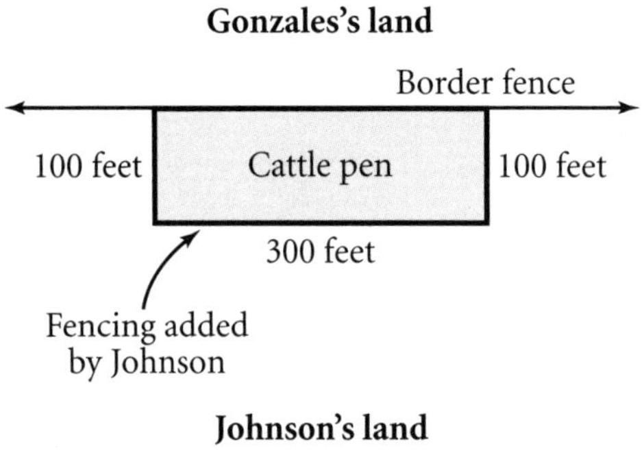

Gonzales's land

Border fence

100 feet | Cattle pen | 100 feet

300 feet

Fencing added by Johnson

Johnson's land

1. What is the area of the cattle pen shown in the diagram?

2. Choose three other possibilities for the dimensions of the rectangular pen. Find the area of each pen. Keep in mind that dairyman Johnson can use as much or as little of the existing border fence as he likes. Also remember that he wants to use a total of 500 feet of fencing for the other three sides.

3. Suppose the pen extends x feet away from the border fence. For example, in the diagram, x would be 100. Find an expression for the area of the pen in terms of x.

4. Try to determine the value of x that will maximize this area.

5. Did you notice anything that this activity has in common with the path of a rocket?

Parabolas and Equations I

In *A Corral Variation,* you might have used a graph (showing area as a function of x) to find the maximum area of the cattle pen. But the graph of the area function was not linear. In the unit problem, the graph of the equation for the rocket's height, $h(t) = 160 + 92t - 16t^2$, is not linear. The general shape of these graphs is called a **parabola.**

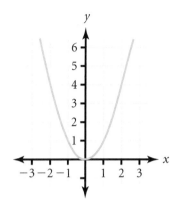

This activity involves equations for parabolas. You will explore the effects that changes in the equations have on their graphs.

1. The simplest parabola equation is $y = x^2$. Enter the equation $y = x^2$ into your calculator. Display its graph. The lowest point on this parabola is called its **vertex.** What are the coordinates of the vertex?

2. The equation $y = ax^2$ is a whole family of graphs. Here are a few members of this family.

$$y = 3x^2 \quad y = 0.5x^2 \quad y = -2x^2 \quad y = 0.001x^2 \quad y = -6x^2$$

Each number value of a produces a slightly different graph. Use your calculator to explore what happens to the graphs of $y = ax^2$ for different values of a.

   a. What's special about graphs that have $a < 0$?

   b. How are graphs that have $a > 1$ different from graphs with $0 < a < 1$?

3. Make a design like each of these graphs on your calculator. Write the equations you use.

a.

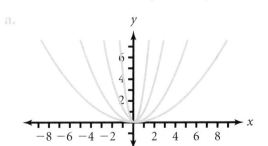

b.

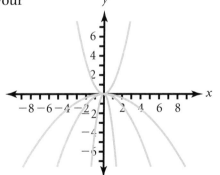

Parabolas and Equations II

1. The equation $y = x^2 + k$ is another family of graphs, where k is a particular number. When k equals 0, you get the simplest member of this family, $y = x^2$.

   a. Write four equations in this family. Graph them all on one screen.

   b. Explore what happens to the graphs of $y = x^2 + k$ for other values of k. How are graphs with $k > 0$ different from graphs with $k < 0$? How is the value of k related to the coordinates of the vertex?

2. Try to make each of these designs on your calculator. Write the equations you use.

   a.
   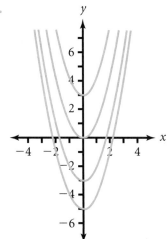

   b.
   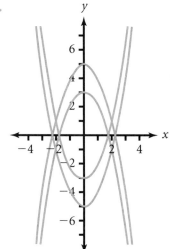

   c.
   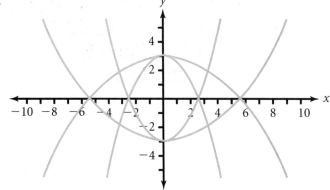

Rats in June

The POW *Growth of Rat Populations* asks you to find how many rats will be on the island on January 1. This is one year after the original female has her first litter.

This activity will get you started on that task. Figure out how many rats will be on the island on June 1, partway through that year.

Parabolas and Equations III

1. The equation $y = (x - h)^2$ is another family of graphs where h is a particular number. What does the value of h have to be to get the simplest member of this family, $y = x^2$?

   a. Write four equations in this family. Graph them all on one screen.

   b. Explore what happens to the graphs of $y = (x - h)^2$ for other values of h. How are graphs with $h > 0$ different from graphs with $h < 0$? How is the value of h related to the coordinates of the vertex?

2. The families $y = ax^2$ and $y = (x - h)^2$ can be generalized together as $y = a(x - h)^2$. Use this form to try to make each of these designs on your calculator. Write the equations you use.

a.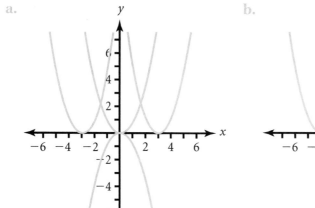

b.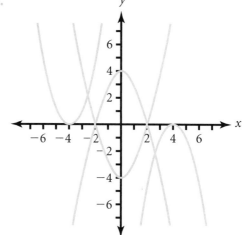

Vertex Form for Parabolas

In *Parabolas and Equations I, II,* and *III,* you explored the equations $y = x^2, y = ax^2, y = x^2 + k,$ and $y = (x - h)^2.$

All these ideas are combined in one general equation, $y = a(x - h)^2 + k.$ This equation is called the **vertex form** for parabolas. The values for $a, h,$ and k are the *parameters* of the equation.

Use what you have learned about these parameters to make each of these designs on your calculator. Write the equations you use.

1.

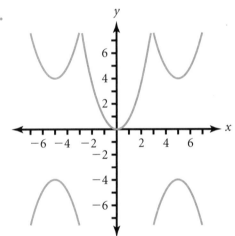

2.

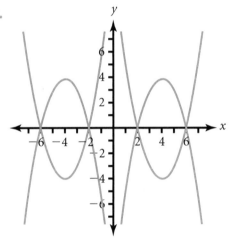

3.

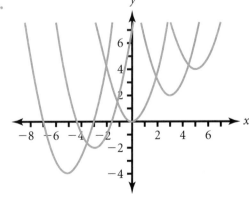

4.

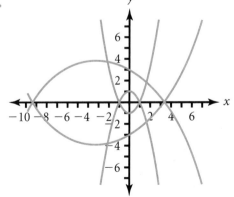

Using Vertex Form

1. Find the equations that will make this design.

2. Can you find the coordinates of the vertex without even graphing the parabola? Find the vertex coordinates first and then check by graphing.

   a. $y = 2(x - 3)^2 + 7$

   b. $y = 0.8(x + 3.5)^2 - 6.2$

   c. $y = -0.3(x - 15)^2 + 20$

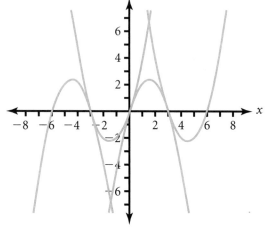

3. The rocket formula given at the beginning of the unit is $h(t) = 160 + 92t - 16t^2$. This formula is not in vertex form. Estimate the vertex of this parabola by graphing.

Crossing the Axis

In some situations, knowing whether the graph of a quadratic function has any x-intercepts is useful.

For each equation, do these steps.

- Determine whether the vertex lies above, below, or on the x-axis.
- Determine whether the parabola is concave down or concave up.
- Use those facts to determine how many x-intercepts the graph has.

1. $y = -(x - 2)^2 + 4$

2. $y = 3(x + 1)^2 + 5$

3. $y = -5(x - 4)^2 - 8$

4. $y = 4(x + 6)^2$

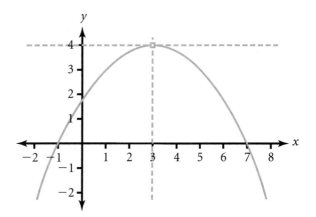

In some situations, you need the equation of a parabola with a particular vertex and a specific **x-intercept.** For example, consider the parabola shown here. It has its vertex at $(3, 4)$ and an x-intercept at $(7, 0)$.

The vertex is at $(3, 4)$, so you know $y = a(x - 3)^2 + 4$. Now, what value of a will make this parabola go through $(7, 0)$? One way to find a is to substitute the value 7 for x and 0 for y into the equation to get $0 = a(7 - 3)^2 + 4$. Copy this equation and solve it for a. Now you can write the equation for the parabola. Check your answer by graphing.

Find an equation for each of these parabolas.

5. Vertex at $(4, -2)$ and an x-intercept at $(7, 0)$

6. Vertex at $(20, 10)$ and an x-intercept at $(40, 0)$

Is It a Homer?

When a baseball is thrown or hit, its path through the air is almost a perfect parabola. Mighty Casey hit a towering drive to center field. The ball reached its maximum height of 80 feet. At this point, the ball is right above a spot on the ground exactly 200 feet from home plate.

The center field fence is 380 feet from home plate and is 15 feet tall. Will Casey's hit clear the fence for a home run? Use mathematics to justify your answer.

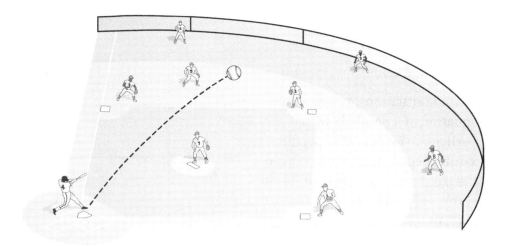

The Form of It All

On a parabola, the vertex is a key point. When the equation of a parabola is in vertex form, $y = a(x - h)^2 + k$, the vertex is easy to find.

In the *Fireworks* unit problem, you want to find when the rocket will reach its maximum height. To get the answer, you will need to find the vertex of the parabola $h(t) = 160 + 92t - 16t^2$.

Unfortunately, the equation $h(t) = 160 + 92t - 16t^2$ is not in vertex form. How can you use algebra to change this equation into vertex form?

You will prepare for this by studying ways to multiply, square, and factor algebraic expressions. You will also develop some new perspectives on the distributive property. Along the way, you will see how to transform a parabolic function from vertex form, $y = a(x - h)^2 + k$, into standard form, $y = ax^2 + bx + c$.

Elana Cohen discusses the "Fireworks" unit problem with a classmate.

A Lot of Changing Sides

A housing developer wants to build a new housing development. She submits plans to the city planner for the new houses. The lots in the plan are all squares of the same size. The city planner insists that the developer introduce some variety into the plan.

After some discussion, the planner and the developer decide the lots should include other types of rectangles. So the developer changes the lengths of the sides of some of the lots.

All the changes described in Questions 1 to 6 are comparisons with the original square lot. For each question, complete these tasks.

* Make and label a sketch of the new lot. Use the variable X to represent the length of a side of the original square.

* Write an expression for the area of the new lot as a product of the length and width.

* Write an expression without parentheses for the area of the new lot as a sum of smaller areas. Use your sketch to explain your expression.

1. The new lot is 3 meters longer in the north-south direction. It is 4 meters longer in the east-west direction.

2. The new lot is 5 meters longer in the north-south direction. It is the same length in the east-west direction.

3. The new lot is 10 meters longer in the north-south direction. It is 9 meters longer in the east-west direction.

4. The new lot is 1 meter longer in the north-south direction. It is 25 meters longer in the east-west direction.

5. The new lot is 2 meters longer in the north-south direction. It is 3 meters shorter in the east-west direction.

6. The new lot is 3 meters shorter in the north-south direction. It is 4 meters shorter in the east-west direction.

Distributing the Area I

The large rectangle shown here is cut by lines parallel to its sides into four smaller rectangles. In the next few days, you will use this area model to multiply and factor algebraic expressions. Each question below is based on this diagram.

The big idea is that the area of the large rectangle can be found in two ways.

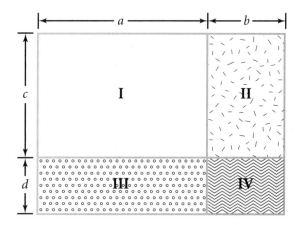

- Multiply length $(a + b)$ by width $(c + d)$.
- Find the area of each of the four smaller rectangles and add them.

1. Suppose $a = 8$, $b = 3$, $c = 5$, and $d = 2$. Draw a diagram and label it with these numbers. Figure out each small area and the total area.

2. Suppose area I $= 28$ square units, $a = 7$, $d = 2$, and area IV $= 6$ square units. Draw a diagram and label it with these numbers. Figure out each missing length and area.

3. Suppose a and c are the same length. Which of the rectangles must be square?

4. Suppose area I is a square with an area of 81 square units. If $b = 4$ and $d = 3$, find all other missing lengths and areas, including the total area.

5. Suppose area I is a square with side length x. If area III is $8x$ square units and area IV is 24 square units, find algebraic expressions for the missing lengths, areas, and total area.

6. Suppose the total area is 864 square units, $a = 30$, and $c = 20$. Find the missing lengths and areas.

Views of the Distributive Property

The **distributive property** is an important principle in mathematics. You can use it in many situations to write an expression in another form. In its simplest algebraic form, the distributive property can be expressed by an equation like this one.

$$a(x + y) = ax + ay$$

In words, you might state the distributive property this way.

Multiplying a sum by something is the same as multiplying each term by that "something" and adding the products.

In this activity, you will look at various ways to think about and use the distributive property.

Multidigit Multiplication

You may not have realized that you are using the distributive property every time you do multiplication that involves more than one-digit numbers. For example, the product $73 \cdot 56$ can be thought of as $(70 + 3) \cdot 56$. Applying the distributive property, you would get $70 \cdot 56 + 3 \cdot 56$. You might write this problem in vertical form.

$$
\begin{array}{r}
73 \\
\times\ 56 \\
\hline
438 \\
3650 \\
\hline
\end{array}
$$

438   (This is $6 \cdot 73$.)

3650   (This is $50 \cdot 73$.)

continued ▸

The products 6 · 73 and 50 · 73 can also be found using the distributive property. To show all the details, you might write the problem like this.

$$
\begin{array}{r}
73 \\
\times\ 56 \\
\hline
18 \\
420 \\
150 \\
\underline{3500}
\end{array}
$$

18   (This is 6 · 3.)
420   (This is 6 · 70.)
150   (This is 50 · 3.)
3500   (This is 50 · 70.)

Each of the numbers 18, 420, 150, and 3500 is called a *partial product*. Writing a multidigit multiplication problem showing all the partial products is sometimes called the *long form*.

In the usual written form of this problem, the partial products 18 and 420 are not shown individually. Instead their sum, 438, is written. Similarly, you omit the partial products 150 and 3500 and simply write their sum, 3650. The numbers 438 and 3650, which are each the sum of two partial products, are sometimes referred to as *partial sums*.

1. Show how to find the product 32 · 94 using the long form. Show all the partial products.

Multiplication with a Diagram

You can illustrate the product 73 · 56 with an area model like this one. Each smaller rectangle represents one of the partial products. The areas of the two rectangles on the top add to the partial sum 3650. The areas of the two rectangles on the bottom add to the partial sum 438.

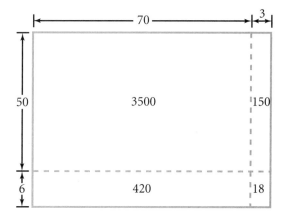

2. Draw a similar area model to illustrate the product 32 · 94.

3. Show how to find the product 47 · 619 in two ways.

continued ▶

Multiplying in Algebra Is Like Multiplying in Arithmetic

Multiplication of algebraic expressions can be done in a way that is similar to multidigit multiplication. For example, you can set up the problem $(2x + 5)(x + 3)$ in vertical form.

$$2x + 5$$
$$\underline{\times\ x + 3}$$

As with the long form of the problem $73 \cdot 56$, this problem involves four separate products.

4. a. Find this product using a vertical-multiplication form. Use the long form or a shorter form.

   b. Find this product using an area model.

Distributing the Area II

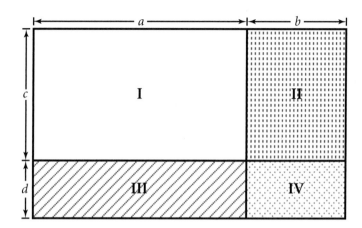

For Questions 1 to 3, use the area model shown here.

1. Suppose the total area is 132 square units, $b = 5$, $d = 4$, and area I is a square. Find the missing lengths and areas.

2. Show how the area model can be used to multiply $(x + 4)$ by $(x + 9)$.

3. Use the area model to multiply $(x + 5)$ by $(x - 2)$.

4. Suppose the total area is the algebraic expression $x^2 + 18x + 80$. What are the algebraic expressions for the length and the width?

Vertical-Multiplication Form

5. Multiply $(x + 5)$ by $(x - 2)$ using the vertical-multiplication form.

6. Multiply $(x^2 + 3x + 2)$ by $(x + 5)$ using the vertical-multiplication form.

7. Multiply the terms in Question 6 using an area model.

Square It!

When a quadratic function is written in the form $y = a(x - h)^2 + k$, you can find the vertex without any computation. This form of the equation is therefore called vertex form. But a typical quadratic function in **standard form** looks like $y = ax^2 + bx + c$, with particular values of a, b, and c. Changing a quadratic from vertex form into standard form involves squaring $(x - h)$, which is written as $(x - h)^2$, and simplifying the result.

This area model shows that $(x + 7)^2 = x^2 + 14x + 49$.

	x	$+$	7
x	x^2		$7x$
$+$			
7	$7x$		49

1. Use an area model or another method to write each expression without parentheses.

   a. $(x + 3)^2$ b. $(x - 5)^2$ c. $(x + 6)^2$ d. $(x - 4)^2$

2. These quadratics are in vertex form. Rewrite each as an equivalent quadratic in standard form.

   a. $y = (x + 5)^2 + 2$ b. $y = 3(x + 5)^2 + 2$ c. $y = 0.5(x - 4)^2 - 5$

3. Find a quadratic function in vertex form that makes each of these graphs. Then rewrite each function in standard form and graph it. How do the graphs compare?

   a.
   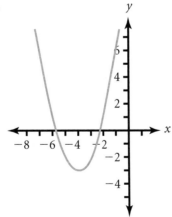

   b.
   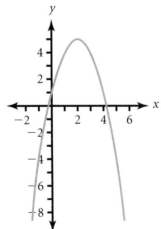

Squares and Expansions

You can square the expression $(x + 6)^2$ to get the **equivalent expression** $x^2 + 12x + 36$. Now you will go in the other direction. This reverse process is called **completing the square.**

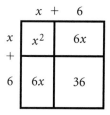

1. Find the number you must add to each given expression so the result is a **perfect square.** Write the result in the form $(x - h)^2$. Draw an area model to illustrate your thinking.

   a. $x^2 - 6x$ b. $x^2 - 5x$ c. $x^2 + 10x$

   d. $x^2 - 4x$ e. $x^2 + 14x$

2. Write each expression without parentheses.

   a. $-(x - 6)^2 + 10$ b. $(x - 2)^2 + 5$

   c. $-(x - 1)^2 + 8$ d. $(x + 3)^2 - 4$

3. Sketch the graph of the function $y = -(x - 5)^2 + 10$. Explain how you used the algebraic representation of the function to make your sketch.

Twin Primes

A prime number is a whole number greater than 1 whose only whole-number divisors are 1 and itself. When two prime numbers differ by exactly 2, the numbers are called twin primes. For instance, 5 and 7, 17 and 19, and 41 and 43 are twin primes.

This activity concerns a curious phenomenon involving twin primes. The phenomenon is true for any pair of twin primes except the combination 3 and 5. If you multiply the twin primes and add 1 to the product, you get a number that has these two properties.

- The number is a perfect square.
- The number is a multiple of 36.

For example, if you start with the twin primes 11 and 13, you get $11 \cdot 13 + 1$, which is 144. This result is equal to 12^2, so this number is a perfect square. It's also equal to $4 \cdot 36$, so it's a multiple of 36. (For the twin primes 3 and 5, you get $3 \cdot 5 + 1$, which is 16. So even in this case, the result is a perfect square, but it isn't a multiple of 36.)

1. Experiment with some other pairs of twin primes. Also try pairs of numbers that are not twin primes. See if you can get some insight into what is happening.

2. Prove the two facts about multiplying twin primes (except 3 and 5) and adding 1.

   - The result is always a perfect square.
   - The result is always a multiple of 36.

   You will need to use a variable in your proof. Think carefully about what number your variable should represent.

Vertex Form to Standard Form

Each of these equations is in vertex form. Write each in standard form, $y = ax^2 + bx + c$, without parentheses. Be sure to combine like terms for your final answer.

1. $y = (x - 1)^2 + 4$
2. $y = -(x - 1)^2 + 4$
3. $y = -(x + 3)^2 + 12$
4. $y = -(x - 3)^2 + 12$
5. $y = -(x + 2)^2 + 7$
6. $y = -(x - 2)^2 + 7$
7. $y = 2(x + 2)^2 + 10$
8. $y = 2(x - 2)^2 + 10$

9. Elsie is making a solar marshmallow cooker out of cardboard and foil. The surface of the cooker will be a parabolic dish, like a TV satellite dish. The dish will focus the sun's rays at a point above the dish. This point is called the focus. It is where Elsie will place the marshmallow for cooking.

To make the cooker, Elsie will cut four parabolic sections out of cardboard. She will cut a notch in the center of each section. Then she will fit the sections together to make a parabolic "skeleton." To make the reflective surface for the cooker, she will cover the skeleton with foil, bright side up.

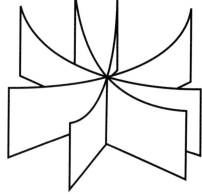

continued ▶

To draw the parabolic cross section, Elsie needs the height values every 10 centimeters, as shown in the diagram below. Figure out how to draw this cross section as a graph on your calculator. Then use your calculator to find the height every 10 centimeters.

Note: Parabolas concentrate incoming energy at a single point called the *focus*. The parabola's capacity to focus energy is why telescope mirrors, satellite TV dishes, headlight reflectors, and solar cookers have a parabolic shape. So where is the focus? Easy. When the equation of a parabola is put into vertex form, $y = a(x - k)^2 + h$, the focal length is exactly $\frac{1}{4a}$ units above the vertex. Elsie won't have to guess where to place her marshmallow. She calculates that when the dish is pointed exactly at the sun, the warmest spot is exactly 40 centimeters above the vertex toward the sun.

How Much Can They Drink?

You may recall farmer Minh from the unit *Do Bees Build It Best?* He built a drinking trough with a triangular cross section for his animals, as shown here. Unfortunately, the trough eventually wore out. He now wants to replace it with one having a rectangular cross section.

He has a metal sheet that is 40 inches wide and 80 inches long. He plans to bend the sheet along two lines parallel to the 80-inch side. This will make a square-cornered U-shape that will form the bottom and the long sides of the trough. He'll use some other pieces of metal for the two ends.

For example, he might mark the sheet with two lines that are 5 inches from each 80-inch edge. These are the dashed lines shown here.

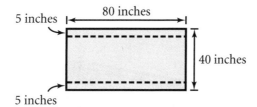

5 inches

80 inches

40 inches

5 inches

Then he would bend up the two 5-inch-wide sections.

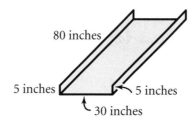

80 inches

5 inches

5 inches

30 inches

continued ▶

Finally, he would attach pieces at the ends to complete the structure. The finished trough might look like the one below. This trough would be 30 inches wide, 80 inches long, and 5 inches high.

Farmer Minh wants to maximize the volume of water the trough will hold. Can you help him?

1. Find the volume of the trough below. It is 30 inches by 80 inches by 5 inches. That is, find out how much water this trough would hold if it were full.

2. Find the volumes for two other troughs farmer Minh could make this way. Use a value other than 5 inches for the width of the sections being bent up.

3. Suppose the sections farmer Minh bends up each have a width of x inches. Find a formula for the volume of the trough.

4. Find the value of x that will make the volume a maximum. Find the volume that goes with that value of x.

5. Suppose that instead of 80 inches, the trough length was increased to 120 inches. How would your answers to Question 4 change?

Putting Quadratics to Use

Question: What do flower beds, medical rescues, widget sales, rockets, and cattle pens have in common?

Answer: They all arise in situations in which quadratic expressions can help you find answers to your questions.

As you explore these diverse contexts, you will strengthen your skills in working with quadratic expressions. You will also learn how to change quadratic functions from standard form into vertex form.

Melody Koker stops to ask if the class has any questions before continuing with her presentation.

Revisiting Leslie's Flowers

In *Leslie's Fertile Flowers,* from *Do Bees Build It Best?,*
landscape architect Leslie needs to know the area of the
triangular flower bed shown below.

One part of finding that area involves figuring out where
the altitude (the dashed line segment) meets the base of
the triangle. In other words, you need to find the value of x.

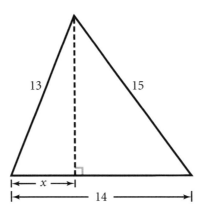

1. Write an equation that will allow you to find x.

2. Solve your equation for x.

Emergency at Sea

A fishing boat has an emergency medical situation. A passenger aboard the boat must be taken to a hospital as soon as possible.

As the diagram shows, there are two lookout towers on shore. They are 1000 meters apart. The technician in tower A determines that the boat is 550 meters from his tower. The technician in tower B finds that the boat is 800 meters from her tower.

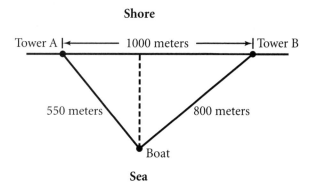

The plan is for the boat to head straight for the nearest point on shore. A medical team will meet the boat there.

1. Show that the point where the boat reaches shore is not 400 meters from tower A.

2. Make one or two other guesses about how far the boat is from tower A when it reaches shore. Determine whether each guess is correct.

3. Set up and solve an equation that will tell how far the boat will be from tower A when it reaches shore.

4. How far is the boat from shore as shown in the diagram?

Here Comes Vertex Form

In the unit problem, the soccer team wants to know the maximum height of the rocket. The team also wants to know how long the rocket will be in the air. You have seen that writing a quadratic function in vertex form, $y = a(x - h)^2 + k$, makes it easy to find the vertex of the graph. The vertex will give you the maximum height of the rocket.

How can you find the time the rocket will be in the air, from the moment it leaves the tower until it hits the ground? You will need to find the x-values for which the height above the ground is zero. In other words, you need to find the x-intercepts of the graph. The vertex form also makes finding these x-intercepts reasonably straightforward.

Remember, the parabola $y = a(x - h)^2 + k$ has its vertex at (h, k).

Complete the square to write each function in vertex form. Then give the coordinates of the vertex. Also state whether the vertex is a maximum or a minimum point.

1. $y = x^2 + 4x + 1$ 2. $y = x^2 - 6x + 15$

3. $y = -x^2 + 8x + 3$ 4. $y = -x^2 - 2x - 5$

The x-intercepts are where the y-value is zero. For each of these quadratic functions, use this idea to write an equation. Solve the equation to find the x-intercepts.

5. $y = (x - 3)^2 - 16$

6. $y = -(x + 1)^2 + 25$

Finding Vertices and Intercepts

For each quadratic function, do these steps.

- Write the function in vertex form.
- Find the vertex of the graph of the function.
- Write an equation and solve it to find the x-intercepts.

1. $y = x^2 + 4x + 3$ 2. $x = x^2 - 6x + 12$

3. $y = -x^2 + 8x + 9$ 4. $y = -x^2 - 2x + 4$

Another Rocket

The Bayside High junior varsity soccer team didn't do as well as the varsity team. The school decided to put on a fireworks display for them anyway. The school used a smaller launching tower, however, so the rocket took off from 80 feet off the ground. And the school used a less-powerful launch mechanism. The rocket rose with an initial speed of only 64 feet per second.

The height of this rocket (in feet) after t seconds is given by the equation $h(t) = 80 + 64t - 16t^2$.

1. Write this height formula in vertex form.

2. Find out these facts about the rocket's travel.

   a. How long did it take for the rocket to reach its maximum height?

   b. What was the rocket's maximum height?

   c. How long did it take for the rocket to hit the ground?

Profiting from Widgets

The Acme Trading Company has acquired 1000 widgets. Its goal is to make as much money from selling them as possible. But the more the company charges for a widget, the fewer it will be able to sell.

Acme's marketing director thinks that if Acme charges d dollars for each widget, the company will be able to sell $1000 - 5d$ widgets. The company will give the rest of the widgets to charity.

1. Suppose Acme charges $70 for each widget.
   a. How many widgets will Acme sell?
   b. How much money will Acme make from sales of widgets?

2. Write an equation that shows how much Acme will get in sales revenue as a function of d.

3. What price per widget will give Acme the maximum possible revenue? How many widgets will the company sell at that price?

4. What do you think about the marketing director's formula?

Pens and Corrals in Vertex Form

1. In *A Corral Variation,* dairyman Johnson wants to build a rectangular cattle pen. He has 500 feet of fencing. He wants to use the existing fence between his property and that of rancher Gonzales as one side of the pen.

   Gonzales's land

   Border fence

   x feet | Cattle pen | *x* feet

   |←——500 − 2x feet——→|

   Fencing added by Johnson

   Johnson's land

   The diagram shows the general situation. The variable *x* represents the distance the pen would extend from the existing fence. Dairyman Johnson needs to choose the value of *x* that will give the pen the greatest possible area.

   a. Develop an expression for the area of the pen in terms of *x*.

   b. Write your expression in vertex form.

   c. Show how to use the vertex form to determine the value of *x* that maximizes the area. Then find that maximum area.

2. In *Do Bees Build It Best?* you saw that of all rectangles with a given perimeter, a square has the greatest area. Now you will use vertex form to prove this statement for a special case.

   a. Use *x* to represent the width of a rectangle with perimeter 200 meters. Find an expression for the area in terms of *x*.

   b. Write your expression in vertex form.

   c. Of all rectangles with perimeter 200 meters, a square has the maximum area. Use the vertex form to explain why this is true. Then find that maximum area.

Vertex Form Continued

1. Transform each quadratic function from standard form into vertex form.

   a. $y = x^2 + 8x + 3$ b. $y = x^2 + 18x$ c. $y = x^2 - 9x + 25$

2. Transform each quadratic function from vertex form into standard form.

   a. $y = -2(x - 2)^2 + 4$ b. $y = 3(x + 1)^2 - 4$ c. $y = -3(x - 4)^2$

3. Find a formula for a parabola with the given vertex and x-intercept.

   a. Vertex at $(0, 5)$ and an x-intercept at $(5, 0)$

   b. Vertex at $(1, 5)$ and an x-intercept at $(5, 0)$

   c. Vertex at $(3, -4)$ and an x-intercept at $(0, 0)$

4. Find a formula for a parabola that approximates the arc of the Golden Gate Bridge. Use the middle of the bridge as $(0, 0)$.

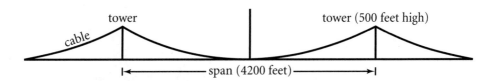

   Use your formula to display the parabola on your calculator. Then find the height of the cable halfway between the midpoint and a tower.

Back to Bayside High

You have acquired several symbolic and graphical tools in this unit. You are now ready to apply them to find precise answers to the questions from the activity *Victory Celebration*.

Jody Lyman, Antonio Cruz, and James Gillum discuss "Victory Celebration".

Fireworks in the Sky

Now you will return to the unit problem described in the activity *Victory Celebration*. As you will recall, the Bayside High varsity soccer team is planning a fireworks display.

The display will use a rocket launched from the top of a tower, 160 feet off the ground. The rocket will initially rise at 92 feet per second.

The rocket's height off the ground (in feet) after t seconds can be found using the function

$$h(t) = 160 + 92t - 16t^2$$

Use the vertex form of this function to find out exactly how long it will take the rocket to reach its maximum height. Then determine what that maximum height is.

You will examine the question about how much time will pass until the rocket hits the ground in the next activity, *Coming Down*.

Coming Down

In *Fireworks in the Sky,* you found when the rocket would reach its highest point and what its maximum height would be.

You now need to determine how long it will take the rocket to return to ground level.

You might be able to get a good approximation using a graph. Your challenge, though, is to find the exact value, using an expression with a square root if necessary.

A Fireworks Summary

You have answered all of the soccer team's questions from the activity *Victory Celebration*.

Your task now is to summarize your work on this problem. Do these tasks as part of your summary.

- Describe the situation from *Victory Celebration* in your own words.
- State the soccer team's questions.
- Carefully explain how you found the answers to those questions. Also give those answers.
- State clearly what those answers tell you about the graph of the height function for the rocket.

Intercepts and Factoring

To solve the unit problem, you had to find the maximum value for a quadratic function. To do so, you learned about the vertex form of quadratic expressions. You also used vertex form to find x-intercepts for quadratic graphs. These are points where the graph meets the x-axis.

Now you will look at another way to find x-intercepts as well as points where the graph intersects other horizontal lines.

Finding these points involves solving quadratic equations. The vertex form will allow you to solve any quadratic equation that has solutions—including those with answers that require square roots. But using vertex form isn't always the simplest approach. In the final activities of this unit, you will see that **factoring** is convenient for solving some quadratic equations.

Alida Jekabson finds a point of intersection.

Factoring

The quadratic function $y = (x + 2)(x - 3)$ is written in factored form as a product of two linear expressions. Factored form is useful for finding x-intercepts. The x-intercepts are the values of x that make $y = 0$.

What values of x will make the product $(x + 2)(x - 3)$ equal to zero? If -2 is substituted for x, you get $(-2 + 2)(-2 - 3) = 0(-5) = 0$. If 3 is substituted for x, you get $(3 + 2)(3 - 3) = 5(0) = 0$.

The function $y = (x + 2)(x - 3)$ in standard form is $y = x^2 - x - 6$. So the x-intercepts for the function $y = x^2 - x - 6$ are at $(-2, 0)$ and $(3, 0)$.

To find the x-intercepts of a quadratic in standard form, $y = ax^2 + bx + c$, you can rewrite the function in factored form.

Rewrite each quadratic function in factored form. An area model, like the one shown in Question 1, may be helpful.

1. $y = x^2 + 6x + 8$

2. $y = x^2 + 9x + 18$

3. $y = x^2 + 11x + 18$

4. $y = x^2 - 10x + 24$

5. $y = x^2 + 10x - 24$

	x	?
x	x^2	$?x$
?	$?x$	8

Let's Factor!

1. Try to write each quadratic expression as a product of two linear expressions. That is, try to write each expression in factored form.

   a. $x^2 + 5x + 6$ b. $x^2 + 2x - 15$

   c. $x^2 + 6x + 10$ d. $x^2 - 9x + 8$

   e. $x^2 - 16$ f. $x^2 - 10x + 6$

2. Find the x-intercepts, if any, for each quadratic function.

   a. $y = x^2 + 5x + 6$ b. $y = x^2 + 2x - 15$

   c. $y = x^2 + 6x + 10$ d. $y = x^2 - 9x + 8$

   e. $y = x^2 - 16$ f. $y = x^2 - 10x + 6$

Solve That Quadratic!

Finding the x-intercepts of a quadratic function $y = x^2 + bx + c$ is basically the same as solving the quadratic equation $x^2 + bx + c = 0$.

For example, the function $y = x^2 - 5x - 6$ has x-intercepts at $(6, 0)$ and $(-1, 0)$. So the solutions to the equation $x^2 - 5x - 6 = 0$ are the values 6 and -1.

1. Find the solutions to each quadratic equation.

   a. $x^2 + 7x + 6 = 0$ b. $x^2 - 3x - 10 = 0$

   c. $2x^2 - 8x + 6 = 0$ d. $x^2 + 4x + 6 = 0$

2. Dairyman Johnson has decided having a pen with the maximum area might not be best. After all, the bigger the pen, the more work it will be to clean it.

   He now thinks that 20,000 square feet is the best possible area for a cattle pen. He is still using a total of 500 feet of fencing, in addition to the fence along the border.

   a. Write an equation whose solution will give the distance the pen should extend out from the existing fence.

   b. Solve your equation and explain the results.

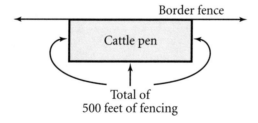

Border fence

Cattle pen

Total of
500 feet of fencing

Quadratic Choices

You have worked with two main methods for finding the x-intercepts for quadratic functions.

- Using vertex form
- Using factored form

For each function, find the x-intercepts using the method you prefer. Explain why you chose that method.

1. $y = -(x + 3)^2 + 25$

2. $y = (x + 9)(x - 6)$

3. $y = x^2 + 12x + 20$

4. The choice of method is hardest when the function is given in standard form, $y = ax^2 + bx + c$. Explain what characteristics of the coefficients a, b, and c might make you choose one method over the other.

A Quadratic Summary

Summarize what you've learned about quadratic expressions, quadratic functions, and quadratic equations. Include ideas about the graphs of quadratic functions.

In your summary, define the terms you use. Here are some ideas. Illustrate your ideas with specific examples. Also give examples of the algebraic techniques you have learned.

Quadratic functions
Standard form
Vertex form
Factored form
Parabola
Vertex
Intercepts
Quadratic equations

Fireworks Portfolio

Now you will put together your portfolio for *Fireworks*. This process has three steps.

- Write a cover letter that summarizes the unit.
- Choose papers to include from your work in the unit.
- Discuss your personal growth during the unit.

Cover Letter

Look back over *Fireworks*. Describe the unit's central problem and main mathematical ideas. Your description should give an overview of how the key ideas were developed. You should also tell how these ideas were used to solve the central problem.

As part of compiling your portfolio, you will select some activities that you think were important in developing the unit's key ideas. Your cover letter should explain why you selected each item.

continued ▶

Selecting Papers

Your portfolio for *Fireworks* should contain these items.

- One or two activities that helped you understand the value of vertex form in solving real-world problems
- One or two activities that helped you become comfortable with the mechanics of working with quadratic expressions
- *A Fireworks Summary*
- *A Quadratic Summary*
- One of the two POWs you completed during this unit: *Growth of Rat Populations* or *Twin Primes*

What About All That Algebra?

This unit included considerable work with the mechanics of algebra. You transformed many algebraic expressions from one form into another.

As part of your portfolio, write about your comfort level with these mechanics. Also discuss the degree to which you think these mechanics are useful or meaningful to you.

A Summary of Quadratics and Other Polynomials

The activities that begin this unit use expressions that involve the square of the variable.

- In *Victory Celebration*, the height of the rocket after t seconds is given by the expression $160 + 92t - 16t^2$.

- In *A Corral Variation*, the area of dairyman Johnson's pen can be described with the expression $x(500 - 2x)$. You can also write this as $-2x^2 + 500x$.

Quadratic Expressions, Functions, and Equations

Both $160 + 92t - 16t^2$ and $x(500 - 2x)$ are quadratic expressions. If the variable is x, the standard form for a quadratic expression is $ax^2 + bx + c$. The variables a, b, and c are specific numbers. The standard form for dairyman Johnson's area expression is $-2x^2 + 500x + 0$. For this example, $a = -2$, $b = 500$, and $c = 0$. The height expression in *Victory Celebration* uses the variable t. Its standard form is $-16t^2 + 92t + 160$, so $a = -16$, $b = 92$, and $c = 160$.

- The variable a is called the coefficient of the quadratic **term.** This coefficient can't be zero. If it were, the expression would be linear rather than quadratic.

- The variable b is called the coefficient of the linear term.

- The variable c is called the **constant term.** The constant term is also considered one of the coefficients in the expression, even though it isn't multiplied by a variable. In dairyman Johnson's area expression, the constant term is zero.

Whenever you have a quadratic expression, you can get an associated quadratic function. The height function $h(t) = 160 + 92t - 16t^2$ is the quadratic function associated with the quadratic expression $160 + 92t - 16t^2$. The coefficients of the expression are also the coefficients of the function.

continued ▶

To find the zeros, or **roots,** of a quadratic expression, set the expression equal to zero. For the expression $2x^2 + 5x - 7$, the associated quadratic equation is $2x^2 + 5x - 7 = 0$.

Graphs of Quadratic Functions

The graphs of quadratic functions all have roughly the same shape. This shape is called a parabola. These graphs can be divided into two categories: those that open upward and those that open downward. A quadratic graph that opens upward has a minimum point. A quadratic graph that opens downward has a maximum point. This special point, whether a minimum or a maximum, is called the vertex. The graphs illustrate the two cases.

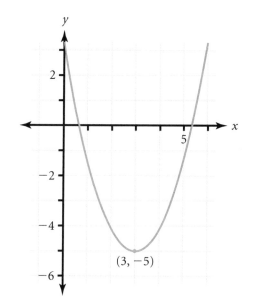

The first graph is of the function $y = x^2 - 6x + 4$. It opens upward, and its vertex is a minimum point at $(3, -5)$.

The second graph is of the function $y = -2x^2 + 4x + 5$. It opens downward, and its vertex is a maximum point at $(1, 7)$.

Polynomials

A quadratic expression is a special case of a polynomial. A **polynomial** in a particular variable is an expression that is a sum of terms. Each term in the sum is either a constant term or a coefficient times a power of the variable (with a positive integer **exponent**).

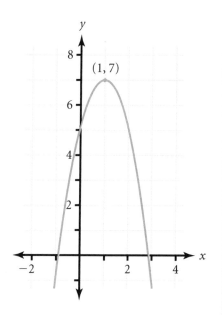

A polynomial with a single nonzero term is called a **monomial.** A polynomial with two nonzero terms is a **binomial.** A polynomial with three nonzero terms is a **trinomial.**

- $4x^2$, $-2y^5$, and 7 are monomials.
- $5t^3 - 8t$ and $z^2 + 9$ are binomials.

continued ▶

- The expression $x(500 - 2x)$ is also a binomial because it is equivalent to the standard quadratic expression $-2x^2 + 500x + 0$, which has two nonzero terms.
- $160 + 92t - 16t^2$ is a trinomial.

Degree, Leading Term, and Leading Coefficient

The highest power of the variable in a polynomial expression is called the **degree** of the polynomial. Thus, all linear expressions are polynomials of degree 1. All quadratic expressions are polynomials of degree 2. A polynomial of degree 3 is called a **cubic** polynomial.

A polynomial that is simply a constant term is a special case. If the constant term is not zero, the degree of the polynomial is 0. The number 0 by itself, considered as a polynomial, is not assigned a degree.

The term of a polynomial involving the highest power of the variable is called the **leading term.** The coefficient in this term is called the leading coefficient. (In the case of a polynomial that is a constant term, that term is both the leading term and the leading coefficient.) Here are some examples.

Polynomial	Degree	Leading term	Leading coefficient
$6t^2$	2	$6t^2$	6
$8w - 2$	1	$8w$	8
$5z^3 - z^2 + 9z + 7$	3	$5z^3$	5
$x^6 + 4x^2$	6	x^6	1
23	0	23	23
0	none	0	0

Sometimes including the "missing" terms—terms whose coefficients are zero—is helpful. For example, you might write the polynomial $x^6 + 4x^2$ as $x^6 + 0x^5 + 0x^4 + 0x^3 + 4x^2 + 0x + 0$. However, you would still consider it a binomial.

SUPPLEMENTAL ACTIVITIES

Most of the supplemental activities for *Fireworks* continue to focus on quadratic functions and equations. Some have a broader algebraic focus, however. Here are some examples.

- *Choosing Your Intercepts* involves finding a quadratic function with a specific vertex and *x*-intercepts.

- *Equilateral Efficiency* is a challenging problem that applies ideas from this unit to prove a geometry principle used in the unit *Do Bees Build It Best?*

- *Factors of Research* gives you an opportunity to learn more about factoring.

- *Standard Form, Factored Form, Vertex Form* provides extra practice in transforming these quadratic forms.

What About One?

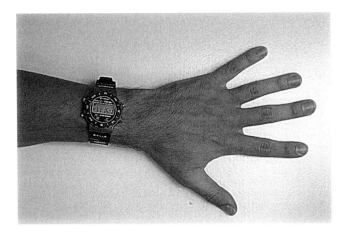

Most digital watches show the hour, the minute, and the second in a format like the one shown here. Your question in this activity is

What fraction of the time is at least one of the digits a 1?

To begin with, assume the watch is in 12-hour mode. For example, two hours after the hour of noon is shown as 2:00, not 14:00.

When you answer the question, consider some or all of these variations.

- What if the watch is in 24-hour mode?
- What if you include a date on the watch (using a format such as 10-17 for October 17)? Assume that it is not a leap year.
- What if the watch shows only the hour and the minute?

Quadratic Symmetry

You have seen that the graphs of quadratic functions are symmetrical. Specifically, the vertical line through the vertex forms a line of symmetry. For every point on the graph to the right of this line, there is a matching point an equal distance to the left of this line. The two points have the same y-coordinate.

Your main task in this activity is to prove this fact using algebra.

1. Consider a quadratic function defined by the equation $y = a(x - h)^2 + k$. Its vertex is the point (h, k).

   a. What is the equation of the vertical line through the vertex?

   b. Suppose a point on the graph is t units to the right of the line of symmetry. What is its x-coordinate?

   c. Suppose a point on the graph is t units to the left of the line of symmetry. What is its x-coordinate?

   d. Show that these two x-values give the same y-value when substituted into the equation $y = a(x - h)^2 + k$.

2. Explain how your work in Question 1 proves the graph is symmetric.

3. Suppose you know that the points $(3, 7)$ and $(10, 7)$ lie on the graph of a particular quadratic function. What can you conclude about the vertex of this graph?

4. Generalize from Question 3. Suppose the points (u, w) and (v, w) lie on the graph of a quadratic function. What can you conclude about the vertex of this graph?

Subtracting Some Sums

Remember the chefs from the Year 1 unit *Patterns* who used hot and cold cubes to change the temperature of their cauldron? You may want to use that model to help you think about these problems.

1. Write each expression as an equivalent expression without parentheses. Simplify your results where possible by combining like terms.

   a. $35 - (3a + 14)$ b. $50 - (c + 17 + 2d)$

   c. $16 + 9s - (3s + 19)$ d. $6 + 4t - (3 + 7t)$

2. In these expressions, you need to multiply the expression in parentheses by a constant and then subtract the whole result. Write each expression as an equivalent expression without parentheses, and simplify by combining like terms.

   a. $5w + 23 - 2(w + 7)$ b. $5 + 6x - 3(x + 1)$

   c. $14 + 7h - 4(h + 5)$ d. $4y + 9 - 2(3y + 4)$

Subtracting Some Differences

Remember the chefs from the Year 1 unit *Patterns* who used hot and cold cubes to change the temperature of their cauldron? You may want to use that model to help you think about these problems.

1. Write each expression as an equivalent expression without parentheses. Simplify your results where possible by combining like terms.

   a. $35 - (3a - 14)$ b. $50 - (c - 17)$

   c. $16 + 9s - (3s - 11)$ d. $6 + 4t - (3 - 7t)$

2. In these expressions, you need to multiply the expression in parentheses by a constant and then subtract the whole result. Write each expression as an equivalent expression without parentheses, and simplify by combining like terms.

   a. $5w + 23 - 2(w - 7)$ b. $5 + 6x - 3(x - 1)$

   c. $14 + 7h - 4(h - 5)$ d. $4y + 9 - 2(3y - 4)$

Choosing Your Intercepts

Suppose you want to find a quadratic function whose vertex is at the point $(2, 9)$. Four such functions are shown on the graph.

Find a quadratic function with vertex at $(2, 9)$ and the given x-intercepts. Or explain why it's mathematically impossible to do so.

1. x-intercepts at $(-1, 0)$ and $(5, 0)$

2. x-intercepts at $(1, 0)$ and $(3, 0)$

3. x-intercepts at $(0.5, 0)$ and $(3.5, 0)$

4. x-intercepts at $(-2, 0)$ and $(7, 0)$

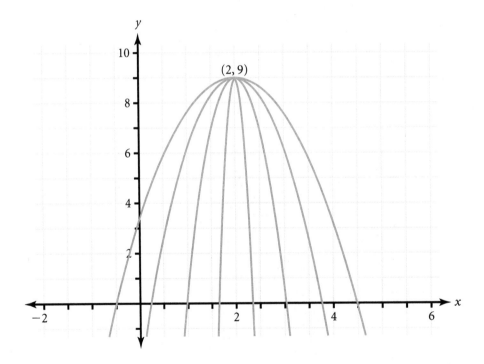

A Lot of Symmetry

The housing developer from *A Lot of Changing Sides* likes symmetry very much. The city planner, though, will not allow square lots. So the developer wants to do something symmetric with rectangular lots.

She decides to see what happens if she starts with square lots, increases the width by some amount, and decreases the length by the same amount. For example, if she increases the length by 4 meters and decreases the width by 4 meters, the area can be represented as $(x + 4)(x - 4)$.

The developer discovers that an interesting pattern emerges if she multiplies out expressions like this and writes them in standard form.

Find the pattern and describe it fully. Then justify the pattern using both algebra and a diagram.

Divisor Counting

As you saw in the POW *Twin Primes,* a prime number is a whole number with exactly two whole-number divisors. This activity is about counting the divisors for *any* whole number (not simply primes). Throughout this activity, the word *divisor* will mean whole-number divisor.

The number 1 is a divisor of every whole number. Every whole number is a divisor of itself. Therefore, every whole number greater than 1 has at least two distinct divisors and so must either be prime or have more than two divisors.

Your task is to figure out as much as you can about how many divisors a number has. You will probably find the concept of prime numbers useful, both in conducting your investigation and in stating your conclusions.

Here are some examples of questions to explore.

- What kinds of numbers have exactly three divisors? Exactly four? And so on?

- Do larger numbers necessarily have more divisors?

- Is there a way to figure out how many divisors 1,000,000 (one million) has without actually listing and counting them? How about 1,000,000,000 (one billion)?

- What's the smallest number with 20 divisors?

In addition to answering these questions, come up with your own questions. Also describe any generalizations you discover.

The Locker Problem

On the first day of school, Lana walks down the hallway past a row of lockers. The lockers are numbered from 1 to 100. When Lana arrives, all the lockers are open. Absentmindedly, she closes all the even-numbered lockers—the multiples of 2—as she walks by.

A few minutes later, Jeremy comes by. He decides to touch only lockers whose numbers are multiples of 3. If one of these lockers is open when he goes by, he closes it. If it's closed, he opens it. For example, Lana left locker 3 open, so Jeremy closes it. Lana closed locker 6, so Jeremy opens it.

Then another student comes by and changes the doors (opens or closes them) on all lockers whose numbers are multiples of 4. Then another student changes the doors on lockers whose numbers are multiples of 5. This continues until, finally, a student comes by who changes only locker 100.

Here is the question.

Which lockers are open at the end of this process?

Your task is to determine which lockers end up open and to find an explanation for the result.

Once you're done, explain which lockers would end up open if the locker numbers went up to 1000. Assume the last student changes only locker 1000.

Equilateral Efficiency

In the unit *Do Bees Build It Best?* you saw that of all rectangles with a given perimeter, the square has the greatest area. It then seems reasonable to expect this result might generalize to other regular polygons.

In fact, for any given number of sides and a fixed perimeter, the regular polygon with that number of sides has the greatest area.

In this activity, you will explore why this is true for triangles. You will use Hero's formula, which gives a triangle's area in terms of the lengths of its sides. Hero's formula says that if a triangle has sides of lengths a, b, and c, then its area is given by the equation

$$A = \sqrt{s(s-a)(s-b)(s-c)}$$

where $s = \frac{a+b+c}{2}$. The number represented by s is called the *semiperimeter* of the triangle. The semiperimeter is equal to half the perimeter.

1. Consider an equilateral triangle with a perimeter of 300 feet.

   a. Use Hero's formula to calculate the area of this triangle. That is, substitute 100 for a, b, and c, and 150 for s, into the equation.

   b. Explain another method for finding the area of this triangle. Show that the two methods give the same answer.

   continued ▶

2. Use Hero's formula to find the areas of other triangles with a perimeter of 300 feet. That is, try different combinations of values for a, b, and c with a sum of 300. Use Hero's formula to find the area in each case. Look for patterns or general principles that describe your results.

The rest of this activity involves proving that the equilateral triangle is the most "efficient."

3. Continue with the case of a triangle whose perimeter is 300 feet.

   a. Suppose a is equal to 80. Use Hero's formula to find an expression for the area in terms of just b. In this case, $b + c$ must equal 220. Find an expression for c in terms of b and then express the area in terms of just b. Remember that s is still equal to 150 because the perimeter is still 300.

   b. Prove that your area expression in terms of b has its maximum when $b = 110$. This area expression should be the square root of some quadratic expression. All you need to show is that this quadratic expression is a maximum when $b = 110$.

   c. Repeat the approach in parts a and b using other values for a. Then generalize your results. That is, if you know a, how should you choose b and c (in terms of a) to maximize the area?

   d. Develop an expression that gives the maximum possible area for a given value of a. Use your result from part c.

   e. Find the value of a that maximizes your expression in part d. (Do not expect to be able to prove algebraically that your answer is the maximum. That requires calculus.)

   f. Explain what your answer from part e has to do with the idea of "equilateral efficiency."

Check It Out!

Jenna and her friend were reviewing ideas about how to solve equations. One basic principle seemed to involve doing the same thing to both sides of an equation.

Jenna was applying this idea to a problem her friend had made up when something strange happened. Here is the problem she was working on.

$$\sqrt{2x - 3} = -5$$

She didn't want a square root in the problem. So she squared both sides of the equation.

$$\left(\sqrt{2x - 3}\right)^2 = (-5)^2$$

She then simplified both sides of this equation.

$$2x - 3 = 25$$

Then she proceeded as usual.

$$2x - 3 + 3 = 25 + 3$$

$$2x = 28$$

$$\frac{2x}{2} = \frac{28}{2}$$

$$x = 14$$

The trouble began when she substituted her answer back into the original equation. Substituting 14 for x, she wanted to verify that $\sqrt{2 \cdot 14 - 3}$ equals -5. But when she simplified the expression $\sqrt{2 \cdot 14 - 3}$, she got $\sqrt{25}$, which is 5, not -5.

continued ▶

In other words, $x = 14$ is not a solution to the original equation.

1. Why do you think Jenna's solution did not check? What do you suppose she did wrong?

2. An apparent solution that does not check is called an extraneous solution. Use Jenna's method to solve these equations. See if you can find a rule for determining when an extraneous solution will occur.

    a. $\sqrt{3y - 2} = -7$

    b. $\sqrt{5w + 6} = 9$

    c. $\sqrt{4a + 1} + 12 = 1$

    d. $\sqrt{2c - 3} = 5$

The Quadratic Formula

A useful formula that gives the x-values that solve $ax^2 + bx + c = 0$ is the quadratic formula. The quadratic formula is

$$x = \frac{-b \pm \sqrt{b^2 - 4ac}}{2a}$$

To use the quadratic formula to find the x-intercepts for a quadratic function, first put the function into standard form, $y = ax^2 + bx + c$. Then carefully substitute the values of a, b, and c into the quadratic formula and simplify.

1. Use the quadratic formula to find the x-intercepts for each function. Compare your results to your answers from the original problems.

   a. The function $y = x(500 - 2x)$ from *A Corral Variation*

   b. The function $y = 80 + 64t - 16t^2$ from *Another Rocket*

2. Start with the general quadratic equation $ax^2 + bx + c = 0$ and derive the quadratic formula. You will need to complete the square using the general values a, b, and c.

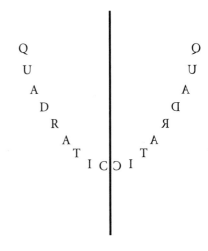

Let's Factor Some More!

Most of the quadratics you factored in this unit had a coefficient of 1 for the x^2 term. Quadratics with coefficients other than 1 can also be factored.

For example, $3x^2 + 14x + 8 = (x + 4)(3x + 2)$, as shown in the area model.

Try to factor each quadratic expression.

	x	$+$	4
$3x$	$3x^2$		$12x$
$+$			
2	$2x$		8

1. $2x^2 + 11x + 5$

2. $2x^2 + 7x + 5$

3. $3x^2 - 7x + 4$

4. $2x^2 + 8x + 9$

Vertex Form Challenge

Putting a quadratic function into vertex form is easiest when the coefficient of x^2 is 1. However, it's possible for *any* quadratic function to be written in vertex form.

Here are some functions in which the coefficient of x^2 is not 1. Write each in vertex form.

1. $y = 2x^2 + 4x + 8$ 2. $y = -2x^2 + 5x + 11$

3. $y = 3x^2 + 6x + 2$ 4. $y = -5x^2 + x + 1$

A Big Enough Corral

1. Dairyman Johnson has decided that he doesn't necessarily need the largest possible corral. However, he wants to be sure his corral has an area of at least 30,000 square feet. The diagram shows the situation.

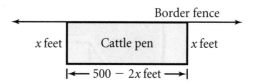

a. Write an inequality that states that the area is big enough for dairyman Johnson. Use the variable x.

b. Sketch the graph of the area function, $y = x(500 - 2x)$. Use scales that make sense for this situation. Be sure the vertex and intercepts show on your graph.

c. Determine the set of values for x that will make the corral big enough for dairyman Johnson's needs. Use either your algebraic inequality or your graph (or perhaps both). Explain your work in detail.

Here's a purely mathematical problem involving a quadratic inequality.

2. Sebastian said, "I'm thinking of a number. If I multiply my number by the number that is 1 less than my number, I get a product that is greater than 12."

a. Write an inequality that expresses this problem.

b. Show that your inequality is equivalent to the inequality $(x + 3)(x - 4) > 0$.

c. Describe the set of numbers that fit your inequality.

Factors of Research

There is more to the topic of factoring than you've seen so far. The questions in this activity suggest some further areas to explore. You also might come up with other questions on your own.

1. Your work has focused on factoring quadratic expressions. What about factoring polynomials other than quadratics? Is it possible? Are there any standard techniques?

2. In *A Lot of Symmetry,* the housing developer from *A Lot of Changing Sides* worked with lots whose areas could be represented by $(x + 4)(x - 4)$. This product can be written as $x^2 - 4^2$, so it is a "difference of squares." What can you say about factoring differences of cubes? Of fourth powers? Of higher powers?

Make Your Own Intercepts

You have used vertex form as a way to find the x-intercepts of quadratic functions. But if you are interested only in x-intercepts— and not in the vertex itself—there may be easier approaches.

1. Write an equation for a function whose graph has exactly two x-intercepts, at $x = 4$ and $x = 2$.

2. Write an equation for a function whose graph has exactly two x-intercepts, at $x = -6$ and $x = 3$.

3. Write an equation for a function whose graph has exactly three x-intercepts, at $x = -5$, $x = 1$, and $x = 5$.

4. Do you think it's possible to create a function with any given set of x-intercepts? Explain your answer.

5. Do you think more than one function fits the condition in Question 1? Can you find another such function? What about the conditions in Question 2 or Question 3?

Quadratic Challenge

1. Solve each quadratic equation. Give your answers both in exact form (using square roots if needed) and as decimal approximations to the nearest tenth.

   a. $x^2 - 6x + 2 = 0$

   b. $x^2 + 4x - 6 = 0$

   c. $2x^2 + 4x - 1 = 0$

2. Sketch the graph of the function associated with each quadratic equation. Also find the x-intercepts and vertex of the graph.

   a. $y = x^2 - 6x + 2$

   b. $y = x^2 + 4x - 6$

   c. $y = 2x^2 + 4x - 1$

Standard Form, Factored Form, Vertex Form

1. How are a quadratic's factored form, x-intercepts, vertex, and vertex form related? Create a table like the one here, and fill in the missing entries.

Standard form	Factored form	x-intercepts	Parabola's vertex	Vertex form
$y = x^2 + 4x - 5$				
$y = x^2 - 7x - 18$				
				$y = (x - 1)^2 - 25$
	$y = (x - 3)(x + 7)$			
	$y = (x + 2)(x - 6)$			
		$(-4, 0)$ and $(6, 0)$		
$y = x^2 + 6x - 16$				
		$(5, 0)$ and $(\ \ )$	$(1, -16)$	
$y = x^2 - 25$				
	$y = (x - 3)^2$			

2. Suppose you are given the x-coordinate of a point on a parabola. How can you find the matching y-coordinate?

3. How is a parabola's factored form related to its x-intercepts?

4. How are the coordinates of a parabola's vertex related to its x-intercepts?

All About Alice

Exponents, Scientific Notation, and Logarithms

All About Alice—Exponents, Scientific Notation, and Logarithms

Who's Alice?

Once upon a time (actually, in 1865), a man wrote a story about the adventures of an imaginary girl named Alice who traveled to a place called Wonderland. The story quickly became a best seller in England and has since been translated into dozens of other languages. The author, Charles Lutwidge Dodgson, used the pen name Lewis Carroll. In addition to his fiction about Alice, he wrote books about mathematical logic. In this unit, one of Alice's adventures is the basis for your exploration of some ideas about numeric operations, graphs, and algebraic formulas.

Alyssa Spediacci and Sam Regan explore new ideas in the Alice unit.

Alice in Wonderland

In 1865, a book was published that would become the most popular children's book in England: *Alice's Adventures in Wonderland.* The author, Lewis Carroll, lived from 1832 to 1898. Lewis Carroll also wrote *Through the Looking-Glass and What Alice Found There,* a sequel to his original story.

The Situation

Read this excerpt from *Alice's Adventures in Wonderland.* Then answer the questions that follow.

[Alice] found a little bottle . . . , and tied round the neck of the bottle was a paper label, with the words "DRINK ME" beautifully printed on it in large letters. . . .

So Alice ventured to taste it, and finding it very nice . . . , she very soon finished it off.

"What a curious feeling!" said Alice. "I must be shutting up like a telescope!"

And so it was indeed: she was now only ten inches high. . . .

Soon her eye fell on a little glass box that was lying under the table: she opened it, and found in it a very small cake, on which the words "EAT ME" were beautifully marked in currants. . . .

So she set to work, and very soon finished off the cake.

"Curiouser and curiouser!" cried Alice. . . . "Now I'm opening out like the largest telescope that ever was! Good-bye, feet!" (for when she looked down at her feet, they seemed to be almost out of sight, they were getting so far off). . . .

Just at this moment her head struck against the roof of the hall.

continued ▶

The Questions

In Lewis Carroll's story, when Alice drinks from the bottle, she grows shorter. When she eats the cake, she grows taller. But Carroll doesn't say how much shorter or how much taller or even how tall Alice was to start with.

Assume for every ounce of cake Alice eats, her height doubles. For every ounce of beverage she drinks, her height is cut in half.

1. What happens to Alice's height if she eats 2 ounces of cake? What happens to her height if she eats 5 ounces?

2. Find a rule for what happens to Alice's height when she eats C ounces of cake. Explain your rule.

3. What happens to Alice's height if she drinks 4 ounces of beverage? What happens to her height if she drinks 6 ounces?

4. Find a rule for what happens to Alice's height when she drinks B ounces of beverage. Explain your rule.

Logic from Lewis Carroll

Lewis Carroll was a mathematician as well as a novelist. One of his special interests was logic, which might be described as the mathematical science of formal reasoning. Logic analyzes how to draw legitimate, or valid, conclusions from true statements. This process of drawing conclusions is called *deduction*.

One of Lewis Carroll's books gives problems involving groups of statements. The reader is supposed to figure out what, if anything, can be deduced from the statements—that is, what new conclusions can be drawn. All of the groups of statements in this POW are adapted from Lewis Carroll's work.

○ *Example 1*

a. John is in the house.

b. Everyone in the house is ill.

If you know that statements a and b are both true, then you can deduce that John must be ill. So "John is ill" is a valid conclusion.

○ *Example 2*

a. Some geraniums are red.

b. All these flowers are red.

In this case, knowing that statements a and b are both true does not tell you whether any or all of "these flowers" are geraniums. They might be other kinds of red flowers. So there isn't anything new you can deduce from these two statements.

○ *Part I: Finding Conclusions*

Examine each set of statements. Decide what, if anything, you could deduce if you knew the statements were true. There may be more than one possible conclusion. Give as many conclusions as you can.

continued

In each case, explain why you think your conclusions are correct or why you think no new conclusions can be drawn. Diagrams or pictures might help you both analyze the statements and explain your reasoning.

1. a. No medicine is nice.
   b. Senna is a medicine.

2. a. All shillings are round.
   b. These coins are round.

3. a. Some pigs are wild.
   b. All pigs are fat.

4. a. Prejudiced persons are untrustworthy.
   b. Some unprejudiced persons are disliked.

5. a. Babies are illogical.
   b. Nobody who is despised can manage a crocodile.
   c. Illogical persons are despised.

6. a. No birds, except ostriches, are 9 feet tall.
   b. There are no birds in this aviary that belong to anyone but me.
   c. No ostrich lives on mince pies.
   d. I have no birds less than 9 feet tall.

○ Part II: Creating Examples

Make up two sets of statements similar to those in Part I. One of your sets should have a valid conclusion. The other should not.

○ Write-up

1. *Process*

2. *Results:* Give your conclusions (with explanations) for each set of statements in Part I. Also give your sets of statements for Part II and explain why each set does or does not have valid conclusions.

3. *Evaluation:* What does this POW have to do with mathematics?

4. *Self-assessment*

The statements in this activity are adapted from *Symbolic Logic and the Game of Logic* by Lewis Carroll (Dover Publications, New York and Berkeley Enterprises, 1958).

Graphing Alice

In the activity *Alice in Wonderland,* you examined what happened to Alice's height in various situations. Now you will look at that information in an organized way.

Choose a suitable scale for each graph you create. The scales of the two axes do not need to be the same.

1. Alice's height changes when she eats the cake. Assume her height doubles for each ounce of cake she eats.

   a. Find out what Alice's height is multiplied by when she eats 1, 2, 3, 4, 5, or 6 ounces of cake.

   b. Make a graph of this information.

2. Alice's height also changes when she drinks the beverage. Assume her height is halved for each ounce she drinks.

   a. Find out what Alice's height is multiplied by when she drinks 1, 2, 3, 4, 5, or 6 ounces of beverage.

   b. Make a graph of this information.

3. Suppose Alice discovers a new kind of cake. This cake triples her height for every ounce she consumes. Answer Question 1 for this new cake.

4. Suppose Alice finds a different kind of beverage. This beverage reduces her height to one-third of its measure for every ounce she consumes. Answer Question 2 for this new beverage.

5. Compare and contrast your graphs for Questions 1 to 4. In general, what do you think is true of these types of graphs?

A Wonderland Lost

The Amazon rain forest is gradually being destroyed by pollution and agricultural and industrial development. For simplicity, suppose that each year, 10% of the remaining forest is destroyed. Assume, also for simplicity, that the present area of the Amazon rain forest is 1,200,000 square miles.

1. a. What will the area of the forest be after 1 year of this destruction process?

    b. What will the area of the forest be after 2 years of this destruction process?

2. Make a graph showing your results from Question 1 and continuing through 5 years of the destruction process. Include the present situation as a point on your graph.

3. Find a rule for how much rain forest will remain after X years. That is, express the area of the rain forest as a function of X.

4. Explain how this situation and graph relate to Alice's situation.

Extending Exponentiation

The fundamental idea in your exploration of Alice's adventure with the strange cake and beverage is that she grows and shrinks exponentially. The definition of exponentiation as repeated multiplication requires that the exponent be a positive whole number. But what if the exponent is zero? Or negative?

In the upcoming activities, you'll use Alice's situation to gain insight into how the operation of exponentiation can be extended to these new types of exponents.

Molly Jansen, Danielle Crisler, and Jillian Clark compare the results they found using the additive law of exponents.

Here Goes Nothing

For this activity, assume the cake Alice eats is base 2 cake, as in the original problem.

1. What would happen to Alice's height if she ate 0 ounces of cake? Specifically, what would her height be multiplied by?

2. In *Graphing Alice,* you made a graph of what Alice's height is multiplied by, as a function of how much base 2 cake she eats. Go back and examine your graph. Explain whether your answer to Question 1 makes sense for that graph.

3. In *Alice in Wonderland,* you developed the general rule that eating C ounces of base 2 cake multiplies Alice's height by 2^C. According to this rule, what should Alice's height be multiplied by if she eats 0 ounces of this cake?

4. What does all this seem to point to as the value of 2^0?

A New Kind of Cake

Part I: Base 5 Cake

Alice has discovered base 5 cake! Each ounce she eats of this new cake multiplies her height by 5.

1. Figure out what Alice's height would be multiplied by if she ate 1, 2, 3, or 4 ounces of this cake.

2. Make a graph of your results. Use x for the number of ounces Alice eats and y for the number her height is multiplied by.

3. a. What would Alice's height be multiplied by if she ate 0 ounces of cake?

    b. Consider your answer to part a in connection with your graph from Question 2. Does your answer seem to fit as a likely value for y when x is 0?

    c. Explain what parts a and b have to do with defining 5^0.

4. Use a pattern approach to explain why it makes sense to say $5^0 = 1$.

Part II: Base or Exponent?

5. Graph the equations $y = 2^x$ and $y = x^2$ on the same set of axes as in Question 2. Plot enough points to get a sense of each function's growth. As x gets larger, which graph has the greater y-value?

Piece After Piece

Alice does not always eat her cake in one sitting. Sometimes she eats a little cake, follows the White Rabbit for a while, and then returns to eat more cake. In this activity, you will investigate what happens to her height when she eats piece after piece of the original base 2 cake.

1. Suppose Alice eats a 3-ounce piece of cake, chases the rabbit, and then eats a 5-ounce piece of cake.

   a. What happens to her height? Express your answer in a way that indicates the two separate changes from the two pieces of cake.

   b. Is the result the same as if she had eaten a single 8-ounce piece of cake? Explain.

2. Make up two more pairs of questions like those in Question 1. Answer your questions, and explain your reasoning.

3. Now answer Questions 1 and 2 as if they were about the beverage instead of the cake.

When Is Nothing Something?

Jen says,

> "The number 0 stands for nothing. So 3^0 means no 3s. No 3s is zero, so 3^0 equals 0."

Ken says,

> "The number 0 stands for nothing. That means 3^0 is the same as a 3 with no exponent, and that's just 3. So 3^0 equals 3."

1. Explain to Jen and Ken why 3^0 is not equal to 0 or 3.

2. Here are two situations in which the number 0 is not nothing.

   - The 0 in 20 is not nothing, or there would be no difference between 2 and 20.

   - A temperature of 0 degrees is not the same as there not being any temperature (whatever that means).

   Describe two more situations in which 0 is not nothing.

3. Mathematicians have decided it makes sense to define 3^0 and expressions like it as equal to 1. This definition fits well with other principles about exponents, so it makes working with exponents logical.

   When people agree to use a word or symbol in a particular way, that agreement is called a *convention*. Think back to other mathematical topics you have studied. Write about another situation in which you think a convention is involved.

Many Meals for Alice

Alice thinks she will be healthier if she eats fewer sweets. She decides to eat a fixed number of ounces of cake each time she sits down for a meal.

Your task is to find out what would happen to her height after different numbers of meals with a given amount of base 2 cake.

1. Suppose Alice eats 3 ounces of cake at each meal. What will her height be multiplied by after two meals? After three meals? After four meals? After M meals?

2. Experiment with different amounts of cake at each meal and different numbers of meals. Use your examples to develop an expression for what Alice's height will be multiplied by after M meals with D ounces of cake at each meal.

3. Would eating 4 ounces of cake at each of six meals be the same as eating 6 ounces of cake at each of four meals? Why or why not?

4. Make up another example like Question 3. Your example should compare two situations in which you switch the number of ounces with the number of meals.

5. Write a general law of exponents that expresses your observations in Questions 3 and 4.

In Search of the Law

You have seen that if you multiply
two exponential expressions with
the same **base,** such as 2^3 and 2^5,
the product is an expression such
as 2^8. The base is the same as before,
and the exponent is the sum of the
exponents from the factors. This
principle is called the **additive
law of exponents.** This law can be
expressed by the general equation

$$A^X \cdot A^Y = A^{X+Y}$$

Actually, there are many laws that relate to exponents. In this activity,
you will investigate other possible laws.

1. Suppose the exponential expressions being multiplied have different
   bases but the same exponent. That is, consider products of the form
   $A^X \cdot B^X$. Look for a general law for how to rewrite the product as a
   single exponential expression. You might want to examine specific
   cases. For example, use the definition of exponentiation to write the
   expression $3^7 \cdot 5^7$ as a product of 3s and 5s.

2. Suppose the two factors have the same base and the same exponent.
   Do you apply the additive law of exponents or do you use your
   answer to Question 1? Look at specific cases to investigate what to
   do with expressions of the form $A^X \cdot A^X$.

3. A common mistake people make with exponents is to multiply the
   base by the exponent instead of raising the base to the power of the
   exponent. For instance, they might say 2^3 is 6 (because $2 \cdot 3 = 6$)
   when the correct answer is 8 (because $2 \cdot 2 \cdot 2 = 8$).

   Are there any pairs of numbers for which this mistake actually gives
   the correct answer? In other words, are there any solutions to the
   equation $A^X = A \cdot X$? If so, what are they?

Having Your Cake and Drinking Too

The additive law of exponents gives you an easy way to calculate what happens to Alice when she eats several pieces of cake. For example, this equation shows the combined effect of a 17-ounce piece of base 2 cake with a 5-ounce piece of that cake.

$$2^{17} \cdot 2^5 = 2^{17+5} \quad \text{or} \quad 2^{22}$$

You also discovered how to combine several servings of the beverage. For example, this equation shows the combined effect of a 4-ounce serving of base 2 beverage with a 9-ounce serving of that beverage.

$$\left(\frac{1}{2}\right)^4 \cdot \left(\frac{1}{2}\right)^9 = \left(\frac{1}{2}\right)^{4+9} \quad \text{or} \quad \left(\frac{1}{2}\right)^{13}$$

Now your goal is to figure out how to determine the effect on Alice's height of combining base 2 cake and base 2 beverage.

1. What is Alice's height multiplied by if she consumes the same number of ounces of cake and beverage? Express your answer by writing an equation using exponential expressions.

continued ▶

2. Write at least five ways to combine eating cake with drinking beverage that will result in Alice being 8 times her original height. That is, find combinations of amounts of cake and beverage for which her original height will be multiplied by 2^3.

3. a. Find several combinations of amounts of cake and beverage that will result in Alice being 32 times her original height.

   b. Find several combinations of amounts of cake and beverage that will result in Alice being 4 times her original height.

4. Look for a pattern in your answers to Questions 2 and 3. Write a general expression for the amount Alice's height is multiplied by if she eats C ounces of cake and drinks B ounces of beverage.

5. What happens to your rule from Question 4 if B is greater than C?

Rallods in Rednow Land

The ruler of Rednow Land has a very wise adviser. The adviser has saved the country in many ways, such as finding counterfeiters of gold coins. The ruler wants to reward this wise person.

The ruler, who loves to play chess, came up with two choices of rewards for the wise adviser.

• *Choice A:* A billion rallods (A rallod is the official coin of Rednow Land.)

• *Choice B:* The amount of money obtained by putting 1 rallod on one square of a chessboard, 2 rallods on the next, 4 on the next, 8 on the next, and so on, until all 64 squares are filled

1. What does your intuition tell you about which choice would be better?

2. Make a decision based on the results of some computation. Explain your decision.

3. A standard chessboard has 64 squares. How many squares are needed to make the value of Choice B as close as possible to the value of Choice A? Explain your reasoning.

Historical note: This activity is an adaptation of a problem that can be traced to Persia in about the seventh century and that may have originated even earlier in India.

Continuing the Pattern

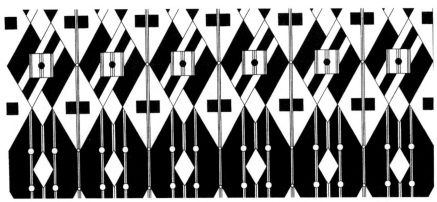

Close-up of an art deco design from "Original Art Deco Allover Patterns," copyright © 1989 by Dover Publications, Inc.

You've seen several ways to explain the definition of 2^0 as 1. Now you will adapt one of those methods to help define exponential expressions involving negative exponents.

1. Begin by examining these powers of 2.

$$2^5 = 32$$
$$2^4 = 16$$
$$2^3 = 8$$
$$2^2 = 4$$
$$2^1 = 2$$
$$2^0 = 1$$
$$2^{-1} = ?$$
$$2^{-2} = ?$$
$$2^{-3} = ?$$
$$2^{-4} = ?$$

a. Look at the equations for powers of 2 with positive and zero exponents. Describe the pattern of values on the right sides of those equations.

b. Explain how you would use this pattern to find the missing values for powers of 2 with negative exponents. Express your numeric results as fractions (not decimals).

continued ▶

2. **a.** Make similar lists for powers of 3 and powers of 5 using positive and zero exponents. Then extend your lists to negative exponents.

   b. Describe how your results for these lists compare with your results for Question 1.

3. For a base of $\frac{1}{2}$, the powers for positive exponents look like this.

$$\left(\frac{1}{2}\right)^5 = \frac{1}{32}$$
$$\left(\frac{1}{2}\right)^4 = \frac{1}{16}$$
$$\left(\frac{1}{2}\right)^3 = \frac{1}{8}$$
$$\left(\frac{1}{2}\right)^2 = \frac{1}{4}$$
$$\left(\frac{1}{2}\right)^1 = \frac{1}{2}$$
$$\left(\frac{1}{2}\right)^0 = 1$$

   a. Extend this list to negative exponents.

   b. Compare your results with your results for Questions 1 and 2.

Negative Reflections

When you first learned about exponents, their use was defined in terms of repeated multiplication. For example, you defined 2^5 as $2 \cdot 2 \cdot 2 \cdot 2 \cdot 2$, using base 2 as a factor five times.

For this repeated-multiplication definition to make sense, the exponent must be a positive whole number. Now you have seen a way to make new definitions that allow zero and negative integers to be exponents.

1. Write a clear explanation summarizing what you have learned about defining expressions that have zero or a negative integer as an exponent. Then explain, using examples, why these definitions make sense. Give as many reasons as you can. Also indicate which explanation makes the most sense to you.

2. Show your explanation to an adult. Ask that person whether it makes sense to him or her. Then write about the person's response.

Curiouser and Curiouser!

Alice remarks, "What a curious feeling!" when she begins shrinking as a result of drinking her special beverage. You might have thought it was rather curious when you learned that zero and negative integers can be used as exponents.

"Curiouser and curiouser!" Alice cries when she first eats the special cake. Your adventures with exponents will also get curiouser and curiouser as you move from integers to fractions in the activities to come.

A Half Ounce of Cake

One day, as Alice is wandering through Wonderland, she stumbles across a silver plate holding a small piece of cake. Alice picks up the cake. She can tell by the size and feel that the cake weighs exactly half an ounce. From the aroma and texture, she knows it is base 2 cake.

1. We all know that eating an ounce of this cake will double Alice's height. But what will eating half an ounce multiply her height by? Keep in mind that eating half an ounce of cake and then eating another half ounce should have the same effect as eating one ounce of cake.

2. Investigate what Alice's height is multiplied by, if she eats other fractional pieces of cake, such as a third of an ounce or a fifth of an ounce.

It's in the Graph

What is $2^{1/2}$?

Maybe you know and maybe you don't. If you don't know, a graph can help you find out. If you do know, a graph will give you another way of thinking about that number.

In *Graphing Alice,* you made a graph showing what Alice's height is multiplied by if she eats various amounts of cake. As you have seen, that graph shows points that fit the equation $y = 2^x$.

In that graph, you considered only positive integers for x. You now know how to interpret the expression 2^x when x is *any* integer.

1. a. Make an In-Out table for the equation $y = 2^x$ using the values $-2, -1, 0, 1,$ and 2 for x. Then plot the points from your table and connect them with a smooth curve.

   b. Use your graph to estimate the value of $2^{1/2}$.

continued ▶

The curve you drew for Question 1 goes through the points (0, 1) and (1, 2). Question 2 deals with the graph of the equation of the *line* through these two points.

2. a. Draw the graph of the equation $y = x + 1$ on the axes you used for Question 1.

   b. Compare the two graphs. What does this comparison tell you about the value of $2^{1/2}$?

3. This question follows a process similar to the one you used in Question 1.

   a. Make an In-Out table for the equation $y = 3^x$, using the values $-2, -1, 0, 1$, and 2 for x. Plot the points and connect them with a smooth curve. Use your graph to estimate the value of $3^{1/2}$.

   b. Use a similar process to estimate the value of $9^{1/2}$ by making a table for the equation $y = 9^x$, plotting and connecting the points, and estimating.

   c. Use a similar process to estimate the value of $\left(\frac{1}{2}\right)^{1/2}$, using the equation $y = \left(\frac{1}{2}\right)^x$.

A Digital Proof

Five boxes are labeled as shown, from Box 0 to Box 4.

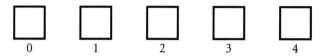

The goal is to put a digit from 0 through 4 *inside* each box so that certain conditions hold.

- The digit you put in Box 0 must be the same as the number of 0s you use.
- The digit you put in Box 1 must be the same as the number of 1s you use.
- The digit you put in Box 2 must be the same as the number of 2s you use and so on.

Of course, you are allowed to use the same digit more than once.

○ What Not to Do

Here is an example of an incorrect way to fill in the boxes.

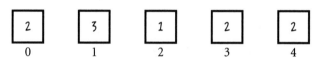

This is incorrect for many reasons. For instance, there is a 1 in Box 2, but more than one 2 is used in the boxes. Similarly, there is a 2 in Box 4, but the number of 4s used is not two.

Your task in this POW is to prove you have found all the solutions.

○ Write-up

1. *Problem Statement*

2. *Process:* Based on your notes, describe how you found all the solutions and how you know you found them all.

3. *Solutions:* List all the solutions you found. Then write a careful and detailed proof that there are no solutions other than those on your list.

4. *Self-assessment*

Stranger Pieces of Cake

In the activity *A Half Ounce of Cake,* you saw how to use Alice's situation to make sense of certain fractional exponents, such as $\frac{1}{2}$ and $\frac{1}{3}$. In these fractions, the numerator is 1.

These are called unit fractions. But what about fractional exponents in general? In this activity, you will investigate the effect of fractional pieces of cake in which the fraction's numerator is not 1.

1. Start with a piece of base 2 cake that weighs $\frac{3}{5}$ of an ounce. What effect should eating this piece have on Alice's height? Explain your answer.

2. Use your work on Question 1 to give a general way of defining $2^{p/q}$ for any fraction $\frac{p}{q}$. Explain your ideas.

Confusion Reigns

The students at Bayside High are learning some fancy stuff about exponents. However, there seems to be some confusion. They are trying to come up with some other generalizations besides the additive law of exponents.

1. Ian and Lian each say they have a way to add expressions with the same base.

    Ian says,

    "$3^4 + 3^5 = 3^9$, because $4 + 5 = 9$."

    Lian says,

    "$3^4 + 3^5 = 6^9$, because $4 + 5 = 9$ and $3 + 3 = 6$."

    Does either student know what's going on? Don't just give yes or no answers. Ian and Lian need clear explanations to help them understand how to work with exponents.

2. Maile, Kylie, and Riley each say they have a way to multiply expressions with the same exponent.

    Maile says,

    "$2^3 \cdot 5^3 = 10^6$, because $2 \cdot 5 = 10$ and $3 + 3 = 6$."

    Kylie says,

    "$2^3 \cdot 5^3 = 10^9$, because $2 \cdot 5 = 10$ and $3 \cdot 3 = 9$."

    Riley says,

    "$2^3 \cdot 5^3 = 10^3$, because $2 \cdot 5 = 10$ and the exponent doesn't change."

    Do any of these students know what's going on? Again, don't simply give yes or no answers. State a particular rule for multiplying expressions with exponents when the exponents are the same, and give clear explanations for your answers.

continued ▶

3. Jen, Ken, and Gwen want to raise exponential expressions to powers.

Jen says,

"$(7^2)^3 = 7^5$, because $2 + 3 = 5$."

Ken says,

"$(7^2)^3 = 7^6$, because $2 \cdot 3 = 6$."

Gwen says,

"$(7^2)^3 = 7^8$, because $2^3 = 8$."

Who's right and who's wrong? State a rule for raising exponential expressions to powers. Give a clear explanation for your answer.

All Roads Lead to Rome

The basic definition for exponential expressions is given in terms of repeated multiplication. For example, 3^5 means "multiply five 3s together." This gives $3 \cdot 3 \cdot 3 \cdot 3 \cdot 3$, or 243. Thus $3^5 = 243$.

In this definition, the exponent tells how many of the bases to multiply together. This definition makes sense when the exponent is a positive integer. But you can't interpret zero, negative, or fractional exponents in terms of how many of the bases to multiply. For example, it doesn't make sense to say, "Multiply negative six 3s together."

You have discovered several ways to make sense of exponents that are not positive integers.

- You can use the Alice story.
- You can extend a numeric pattern that starts with positive integer exponents.
- You can see what definition will be consistent with the additive law of exponents, which says $A^B \cdot A^C = A^{B+C}$.
- You can make a graph of the equation $y = A^x$, using positive integer values for x. You can then use your graph to estimate y for other values of x.

continued ▶

Fortunately, all four approaches lead to the same conclusions. This activity gives you a chance to show how the different methods work.

1. Suppose Alice has base 5 cake and beverage. For each ounce of cake she eats, her height is multiplied by 5. For each ounce of beverage she drinks, her height is multiplied by $\frac{1}{5}$. Explain the meaning of 5^0 using all four methods just described.

2. The four methods don't necessarily all make sense for every possible exponent. Explain the meaning of each of these exponential expressions using as many of the four methods as make sense for that example. You will need to decide in each case what base of cake or beverage to use.

   a. 3^{-4} b. $2^{1/2}$ c. $7^{1/3}$ d. $32^{2/5}$

Measuring Meals for Alice

Alice is sitting down to a meal with her original base 2 cake and beverage. That is, 1 ounce of cake doubles her height and 1 ounce of beverage halves her height.

Find answers to the nearest tenth of an ounce or tenth of a foot for each question. Explain your answers.

1. If Alice is 1 foot tall, how much cake should she eat to become 10 feet tall?

2. a. If Alice is 1 foot tall, how much cake should she eat to become 100 feet tall?

   b. Compare your result in part a to your answer in Question 1. Discuss the connection between the two problems.

3. If Alice is 9 feet tall and wants to be 3 feet tall, how much beverage should she drink?

4. If Alice is 20 feet tall and drinks 2.4 ounces of beverage, how tall will she be?

Turning Exponents Around

If you know what kind of cake and how much cake Alice is eating, you can figure out what her height will be multiplied by. But what if you know only the *kind* of cake and you want her to grow by a certain factor? How can you figure out *how much* cake she should eat?

In these final activities, you will explore questions like this. You'll learn some special ways to express the answers to such questions, as well as a mathematical notation for writing very big and very small numbers.

Hannah Hansel graphs a logarithmic function on her calculator.

Sending Alice to the Moon

1. Alice has just discovered base 10 cake! She is delighted with how powerful it is. One day, after nibbling on the cake, Alice realizes she is 1 mile tall. Having her head in the sky makes her think about space travel. She decides she wants to visit the moon, which is about 239,000 miles from the earth.

   How many more ounces of base 10 cake should Alice eat so the top of her head just touches the moon? Give your answer to the nearest hundredth of an ounce.

2. After her head reaches the moon, Alice continues to enjoy the cake. Eventually her head reaches Pluto, which at the time is approximately 3,670,000,000 miles from the earth. But she forgot to keep track of how many ounces of cake she ate to grow that tall.

   How many ounces of base 10 beverage must Alice drink to return to a height of 1 mile? Give your answer to the nearest hundredth of an ounce.

Alice on a Log

Alice is thinking about base 10 cake and beverage. If she eats 1 ounce of the cake, her height is multiplied by 10. If she drinks 1 ounce of the beverage, her height is multiplied by $\frac{1}{10}$.

Alice has just heard about **logarithms** and is quite excited. For example, she discovered that $\log_{10} 162$ means "the power to which I should raise 10 to get 162."

"How could anyone find that exciting?" you might ask.

Well, Alice is excited because she thinks it sounds more sophisticated to ask, "What is $\log_{10} 162$?" than, "How many ounces of base 10 cake should I eat to grow to 162 times my height?"

1. Between which two whole numbers does the value of $\log_{10} 162$ lie? Explain your answer.

2. For each question, write a logarithm expression that represents the answer. Then find the value of the expression.

   a. How many ounces of base 10 cake should Alice eat to grow to 100 times her height?

   b. How many ounces of base 10 cake should Alice eat to grow to 10,000 times her height?

   c. How many ounces of base 10 cake should Alice eat to grow to 50 times her height?

   d. How many ounces of base 10 cake should Alice eat to grow to 2000 times her height?

   e. How many ounces of base 10 beverage should Alice drink to shrink to $\frac{1}{10}$ of her height?

   f. How many ounces of base 10 beverage should Alice drink to shrink to $\frac{1}{4}$ of her height?

Taking Logs to the Axes

Once Alice discovers that $\log_{10} 162$ means "the power to which I should raise 10 to get 162," she grows curious about what the graph of a logarithmic function looks like. She realizes there is a different logarithmic function for each **base.** For example, one such function is defined by the equation $y = \log_2 x$.

Alice's investigations rely heavily on the fact that the equation $c = \log_a b$ means the same thing as the equation $a^c = b$. This relationship allows her to work with exponential equations, with which she is more comfortable.

1. For each equation, choose values for x for which you can easily compute the value of y. Plot the resulting points. Choose enough points that you can sketch the graph accurately.

   a. $y = \log_2 x$

   b. $y = \log_3 x$

2. Graph the equation $y = \log_{10} x$ on your calculator.

3. Compare the graph of the logarithmic function using base 2 with the graphs of logarithmic functions using other bases. In general, how does the graph change as the base increases? Why?

4. How does the graph of a logarithmic function compare with the graph of the corresponding exponential function?

Base 10 Alice

Alice is enjoying base 10 cake and beverage.

1. Find Alice's height after eating each amount of cake. In each case, assume she starts at a height of 5 feet.

   a. 4 ounces   b. 8 ounces   c. 13 ounces

2. Suppose Alice eats a whole number of ounces of cake and starts at a height of 5 feet. What do you know about the possible heights she can grow to?

3. Suppose Alice is 5 feet tall and wants to know how many ounces of cake she needs to eat to become 50,000,000,000 feet tall. (That's 50 billion feet, or roughly 10 million miles.) What shortcut can you use to answer her question?

4. Pick three whole numbers of ounces of beverage for Alice to drink. Find her height after consuming each amount. Assume in each case that she starts at a height of 5 feet.

5. Find a simple rule for writing Alice's final height for situations like those you made up in Question 4. Your rule should deal specifically with the case of whole-number ounces of beverage.

Warming Up to Scientific Notation

Scientific notation may be a new way for you to express numbers. It often takes some practice to get used to working with scientific notation. It's worth the effort, though, because many ideas in mathematics and science are expressed using this special way of writing numbers.

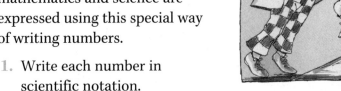

1. Write each number in scientific notation.

   a. 34,200   b. 0.0034

2. Write each number in ordinary notation.

   a. $4.2 \cdot 10^5$ b. $7.503 \cdot 10^{-2}$

3. Each of these problems gives a product or quotient of two numbers written in scientific notation. Find the value of each result *without using a calculator*. Write your answers in scientific notation.

   a. $(3 \cdot 10^4) \cdot (2 \cdot 10^7)$ b. $(5 \cdot 10^5) \cdot (7 \cdot 10^{-2})$

   c. $(7 \cdot 10^8) \div (2 \cdot 10^3)$ d. $(9 \cdot 10^3) \div (3 \cdot 10^{-4})$

   e. $(6 \cdot 10^{-3}) \div (2 \cdot 10^{-8})$ f. $(2 \cdot 10^5) \div (4 \cdot 10^3)$

4. Based on your discoveries in Question 3, develop some general principles for multiplying and dividing numbers written in scientific notation. Make up more examples as needed, and illustrate your rules.

5. a. Figure out and describe how to enter numbers in scientific notation on your calculator.

   b. Describe how your calculator displays numbers in scientific notation.

Big Numbers

Scientific notation can be helpful in working with big numbers.

In some of the questions in this activity, you are given information in scientific notation. Use what you learned in *Warming Up to Scientific Notation* to simplify the computations.

You will probably want to write your answers in scientific notation. But you don't necessarily need to give exact answers. Use your judgment about how much precision is appropriate in each case.

1. A certain computer can do a computation in $5 \cdot 10^{-7}$ seconds. How many computations can the computer do in 30 seconds?

2. A leaking faucet drips 1 drop per second. If there are 76,000 drops of water in a gallon, how many gallons would the faucet drip in a year?

3. Measurements show that Europe and Africa are separating from the Americas at a rate of about 1 inch per year. The continents are now about 4000 miles apart. For simplicity, assume the rate has been constant throughout geologic history. How many years ago did the continents split apart and start drifting?

continued ▶

4. In 2008, the gross national debt of the United States was approximately $9,000,000,000,000. The population at that time was about 305,000,000 citizens. About how many dollars per citizen was the gross national debt in 2008?

5. One atom of carbon weighs approximately $1.99 \cdot 10^{-23}$ gram. How many atoms are in a kilogram of carbon?

6. The earth's mass is about $5.98 \cdot 10^{24}$ kilograms. The sun's mass is about $1.99 \cdot 10^{30}$ kilograms. Approximately how many earths would it take to have the same mass as the sun?

7. Light travels at a speed of approximately 186,000 miles per second. (That's very fast.) A *light-year* is the distance light travels in a year. Approximately how many inches are in a light-year?

8. For simplicity, suppose a grain of sand is a cube that is 0.2 millimeter on each edge. About how many grains of sand packed tightly together would it take to make a beach 300 meters long, 25 meters wide, and 5 meters deep?

Questions 1 to 4 were adapted from *Algebra I* by Paul A. Foerster (Menlo Park, CA: Addison-Wesley Publishing Co., 1990).

An Exponential Portfolio

You have seen that exponents don't have to be positive integers. You have also discovered many general laws about exponents that are based on the definition of exponentiation as repeated multiplication.

Your task now is to list all the general laws of exponents you have studied. Give at least one explanation for each general law. Your explanations may use the Alice story, a numeric pattern, or some other approach.

All About Alice Portfolio

You will now put together your portfolio for *All About Alice*. This process has three steps.

- Write a cover letter that summarizes the unit.
- Choose papers to include from your work in the unit.
- Discuss your personal growth during the unit.

Cover Letter

Look back over *All About Alice* and describe the central theme of the unit and the main mathematical ideas. Your description should give an overview of how the key ideas were developed. Your cover letter should also include an explanation of each item you select.

Selecting Papers

Your portfolio for *All About Alice* should contain

- *Negative Reflections*
- *All Roads Lead to Rome*
- *An Exponential Portfolio*
- An activity in which you used exponents to solve the problem
- An activity that involved graphing
- A Problem of the Week

  Select one of the two POWs you completed during this unit: *Logic from Lewis Carroll* or *A Digital Proof.*

Personal Growth

Your cover letter should describe how the unit develops. In addition, write about your own personal development during this unit. Include any thoughts about your experiences that you wish to share with a reader of your portfolio.

You may also want to address this broader question.

How do you think you have grown mathematically during your second year of IMP?

SUPPLEMENTAL ACTIVITIES

Most of the supplemental activities for *All About Alice* continue the focus on exponents and related ideas. Here are three examples.

- *Inflation, Depreciation, and Alice* explores how changes in prices might involve exponential functions.

- *A Little Shakes a Lot* describes an important use of logarithms.

- *Very Big and Very Small* continues your work with scientific notation.

Inflation, Depreciation, and Alice

As you have probably discovered, prices tend to rise. For example, you may know that it used to cost only 10¢ to use a pay phone, but it now costs 50¢ or $1.00.

This rise in cost is called *inflation*. The inflation rate for an item or service is the percentage increase in its price over time. For instance, an annual inflation rate of 5% means prices go up 5% each year. Of course, the rate of inflation is usually not constant.

1. Suppose in 2008 the price of a large jar of peanut butter was $4.49.

   a. Using an annual inflation rate of 5%, calculate the price of this jar of peanut butter in the year 2025.

   b. Find a rule for computing the price of the peanut butter *N* years after 2008.

In many situations, the value of an item decreases over time. For example, a car that was purchased new for $25,000 five years ago may be worth only $12,000 today. This is called *depreciation*.

If an item depreciates at the rate of 10% per year, the item loses 10% of its value each year. That is, at the end of each year, its value is 10% less than it was at the start of that year.

2. A fitness club purchases a treadmill for $5,800. Because of heavy use, the treadmill depreciates 15% per year.

   a. How much will the treadmill be worth in 10 years?

   b. Find a rule for computing the value of the treadmill after *T* years.

   c. When will the treadmill be worth nothing?

3. What do these problems have to do with *All About Alice?* Describe Alice situations that would fit the rules you found in Questions 1 and 2.

A Logical Collection

This activity has three parts. Each part requires that you use logical reasoning to figure out who is telling the truth.

In your write-up for each part, explain how you solved the problem and how you can prove your answers are correct.

Part I: The Missing Mascot

The mascot of Goldenrod High—a stuffed ostrich—sits outside the main office. Just before the big game, the ostrich disappears. Three students from archrival Greenview High are being questioned.

Each of the students makes some statements.

Adams says,

- "I didn't do it."
- "Benitez was hanging out near Goldenrod that day."

Benitez says,

- "I didn't do it."
- "I've never been inside Goldenrod."
- "Besides, I was out of town all that week."

Clark says,

- "I didn't do it."
- "I saw Adams and Benitez near Goldenrod that day."
- "One of them did it."

Assume two of these students are innocent and are telling the truth and that the remaining student is guilty and may be lying. Who did it? Prove your answer.

continued ▶

Part II: What Did He Say?

You are in a strange place where some people always tell the truth and the rest always lie. You find yourself sitting with three of these people: A, B, and C. You decide to try to determine who belongs to which category.

> You say to A, "Are you a truth-teller or a liar?"
>
> A answers your question, but a squawking bird prevents you from hearing the reply.
>
> B says, "A says he's a truth-teller."
>
> C says, "B is lying."

What can you figure out from this conversation? What can't you figure out? Prove your answers.

Part III: The Turner Triplets

The Turner triplets have a policy that whenever anyone asks them a question, two of them tell the truth and the other one lies. You have just asked them who was born first.

Here are their answers.

> Virna says, "I am not the oldest."
>
> Werner says, "Virna was born first."
>
> Myrna says, "Werner is the oldest."

Who was born first? Prove your answer.

More About Rallods

Part I: Counting Rallods

In *Rallods in Rednow Land,* you studied a situation involving powers of 2. To solve that problem, the wise adviser in Rednow Land might have liked to have an easy way to find the sum of such powers. Perhaps you can help.

1. Find a general formula, in terms of n, for the sum $1 + 2 + 4 + \cdots + 2^n$. You may want to start by investigating some specific examples, choosing small values for n.

Part II: Geometric Sequences

The sequence of powers of 2—that is, 1, 2, 4, 8, 16, and so on—is an example of a geometric sequence. A **geometric sequence** is a sequence of numbers in which each term is a fixed multiple of the previous term. In this example, the multiplier is 2, because each term is twice the one before it.

The multiplier can be any number. For example, if the multiplier is 3 and the first term of the sequence is 1, then the sequence is 1, 3, 9, 27, 81, and so on.

A geometric sequence can have any number as its first term. For example, if the first term is 12 and the multiplier is $\frac{1}{2}$, then the sequence is 12, 6, 3, $\frac{3}{2}$, and so on.

continued ▶

2. Develop a general formula for finding a given term of a geometric sequence if you know the first term and the multiplier. For instance, if the first term is a and the multiplier is r, what is the fourth term? The tenth term? The hundredth term? The nth term?

3. Consider sums of terms of geometric sequences that begin with 1.

   a. Examine sums of the form $1 + 3 + 3^2 + \cdots + 3^k$ for different values of k. Find a formula for such a sum in terms of k.

   b. Look at examples using different multipliers. Try to find formulas similar to those in Question 1 and Question 3a.

   c. Now find a general formula for a sum of the form $1 + r + r^2 + \cdots + r^n$. Your answer should be an expression in terms of r and n.

4. Consider sums for general geometric sequences. Use a to represent the first term and r to represent the multiplier.

   Your goal is to find a general formula, in terms of a, r, and n, for the sum $a + ar + ar^2 + \cdots + ar^n$. Begin by thinking about how such a sum compares with the corresponding sum for the sequence with the same multiplier but whose first term is 1.

Ten Missing Digits

In the POW *A Digital Proof,* you had to place digits in five boxes in a way that satisfied certain conditions and then prove that you found all the solutions. Now you will investigate a more complex version of that problem. This time you don't need to prove you have found the only solution.

There are ten boxes in this puzzle, from Box 0 to Box 9.

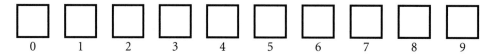

You must put a digit from 0 to 9 in each box so these conditions hold.

- The digit in Box 0 must be the same as the number of 0s you use.

- The digit in Box 1 must be the same as the number of 1s you use.

- The digit in Box 2 must be the same as the number of 2s you use and so on.

You may, of course, use the same digit more than once.

Exponential Graphing

1. Consider the functions f, g, and h defined by these equations.

$$f(x) = 2^{(x^2)}$$

$$g(x) = 2^{2x}$$

$$h(x) = (2x)^2$$

   Investigate the graphs of these functions. Discuss how each graph is the same as or different from the other two graphs.

2. Use an In-Out table to graph the function given by the equation $y = -(2^{-0.5x})$.

3. Explain why the function whose equation is $y = 2^{-x}$ has the same graph as the function whose equation is $y = \left(\frac{1}{2}\right)^x$.

4. Graph the function whose equation is $y = (-3)^x$ using only integer values for x. Describe and explain the behavior of this graph.

Basic Exponential Questions

The numeric value of an expression like a^b depends on both the base and the exponent. In this activity, you will explore some patterns involved in varying the base, the exponent, or both.

Use examples to explain your conclusions.

1. If $n < 0$, which is greater, 2^n or 3^n?

2. If $X < 0$, which is greater, 1^{3X} or 1^{3+X}?

3. Consider the expressions 2^{5m} and 2^{m-1}.
   a. For which values of m is $2^{5m} > 2^{m-1}$?
   b. For which values of m is $2^{5m} < 2^{m-1}$?
   c. For which values of m is $2^{5m} = 2^{m-1}$?

4. If X and Y represent the same number, then the expressions X^Y and Y^X will certainly give the same result. But what if X and Y are not equal? Find out what you can about solutions to the equation $X^Y = Y^X$ in which X and Y are not equal.

Alice's Weights and Measures

When Alice is growing so her head reaches into space, she learns some lessons about approximation.

She decides to visit a star 1 quadrillion miles away. (*One quadrillion* means 10^{15}.) So she adjusts her height to 1 mile and then eats 15 ounces of base 10 cake. She doesn't have a problem eating that much because she is a mile tall to begin with.

Much to her surprise, she is way off her goal when she finishes eating. Then she looks down and discovers some cake crumbs lying at her feet. (Her vision is excellent.)

It turns out she had dropped some of her cake. She asks her friend the Mad Hatter to weigh the crumbs. He discovers she has actually eaten only 14.99 ounces of cake, instead of 15 ounces.

1. Without using a calculator, guess how far Alice is from the star.

2. Now calculate how far Alice really is from the star.

3. What percentage of Alice's 15 ounces of cake did she drop as crumbs?

4. By what percentage of 1 quadrillion miles did Alice miss her goal?

5. Suppose Alice didn't drop any crumbs, but was a little careless when she measured her initial height. Specifically, suppose she was really 0.99 mile tall, rather than 1 mile, and ate exactly 15 ounces of base 10 cake.

    a. Without using a calculator, guess how far Alice is from the star in this case.

    b. Now calculate how far Alice really is from the star.

    c. By what percentage is Alice's initial height measurement off from her estimate of 1 mile?

    d. By what percentage of 1 quadrillion miles does Alice miss her goal?

A Little Shakes a Lot

One of the most familiar uses of logarithms is the Richter scale. This scale offers a numeric way to describe the size of an earthquake. Somewhat simplified, the equation used to compute the Richter scale number of an earthquake is

$$R = \log_{10} a$$

where R is the Richter scale number and a is the amplitude, or amount of ground motion, as measured on a seismograph.

To give intuitive meaning to Richter scale numbers, you need some points of reference. For example, an earthquake that measures 4.0 on the Richter scale is barely perceptible beyond its immediate center. In contrast, the great San Francisco earthquake of 1906, which killed about 3000 people, measured around 7.9 on the Richter scale.

continued ▶

1. The Richter scale numbers make it sound as if the 1906 earthquake was only about twice the size of an earthquake that can hardly be felt. In fact, that isn't the case at all. Prove this by answering these questions.

    a. How many times as much ground motion does an 8.3 quake have compared to a 4.0 quake?

    b. What Richter measurement would represent a quake with twice the ground motion of one that measures 4.0 on the Richter scale?

    c. What Richter measurement would represent a quake with half the ground motion of one that measures 7.9 on the Richter scale?

2. The 2004 Indian Ocean earthquake off the coast of Indonesia and the resulting tsunami left more than 200,000 people dead or missing. This quake measured about 9.3 on the Richter scale. In numeric terms, how did the amount of ground motion in this earthquake compare to that of the 1906 San Francisco quake?

Who's Buried in Grant's Tomb?

There's a silly old riddle that asks, "Who's buried in Grant's tomb?" Based on this riddle, people sometimes use the phrase "Grant's tomb question" to describe a problem that contains its own answer. One example is the question "What was the color of George Washington's white horse?"

1. Here are some questions about exponents and logarithms that might be called Grant's tomb questions. Be sure to explain your answers.

   a. What is the cube root of 17^3?

   b. What is the value of $\log_5 5^8$?

   c. What is the value of $7^{\log_7 83}$?

   d. For what value of x is $\log_x 2^{11}$ equal to 11?

   e. How can you simplify the expression $\left(\sqrt[6]{162}\right)^6$?

2. Make up some Grant's tomb questions of your own. They don't have to involve exponents and logarithms.

By the way, the original Grant's tomb riddle is actually a trick question. The elaborate monument in New York City contains both the body of Ulysses S. Grant—the eighteenth president of the United States—and that of his wife, Julia Grant.

Very Big and Very Small

For this activity, you need to think of two situations in which you would like to investigate numeric questions.

The first situation should involve very large numbers.

The second situation should involve very small numbers.

You can use the examples in *Big Numbers* for ideas.

Write a report on each of your situations. Use reference materials or take actual measurements and estimates to get data for your reports. You may want to check with your teacher about the suitability of your situations and questions.

GLOSSARY

This is the glossary for all five units of IMP Year 2. This glossary may be useful when you encounter a term in **bold** text that is new or unfamiliar, or you can use it to confirm or clarify your understanding of a term.

Absolute growth The growth of a quantity, usually over time, found by subtracting the initial value from the final value. Used in distinction from **percentage growth.**

Additive law of exponents The mathematical principle that states that the equation

$$A^B \cdot A^C = A^{B+C}$$

holds true for all numbers A, B, and C (as long as the expressions are defined).

Altitude (of a **parallelogram, trapezoid** or **triangle**) For a **parallelogram** or **trapezoid,** a line segment (or the length of a line segment) connecting two parallel sides and perpendicular to these two sides. Each of the two parallel sides is called a **base** of the figure.

Examples: Line segment *KL* is an altitude of parallelogram *GHIJ*, with bases $\overline{GJ}$ and $\overline{HI}$. Line segment *VW* is an altitude of trapezoid *RSTU*, with bases $\overline{RU}$ and $\overline{ST}$.

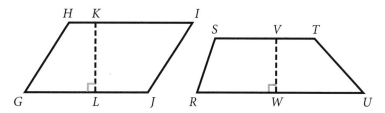

For a triangle, a line segment (or the length of a line segment) from any of the three vertices and perpendicular to the opposite side or to an extension of that side. The side to which the perpendicular segment is drawn is the **base** of the triangle.

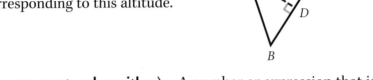

Example: Line segment *AD* is an altitude of triangle *ABC*. Line segment *BC* is the base corresponding to this altitude.

Base (of an **exponent** or **logarithm**)   A number or expression that is raised to a power.

Examples: In the expression 2^8, 2 is the base; in the expression x^4, *x* is the base; in the equation $\log_2 32 = 5$, 2 is the base.

Base (of a **triangle, parallelogram,** or **trapezoid**)   The side of a triangle, a parallelogram, or a trapezoid to which an altitude is drawn.

Base (of a **prism**)   The initial or final position of the polygon that was moved through space to form a prism. See *The World of Prisms* in the unit *Do Bees Build It Best?*

Bias   A systematic overrepresentation of some subgroup of a population, not necessarily as a result of intentional or conscious favoritism.

Binomial   A **polynomial** with exactly two **terms.**

Chi-square statistic   A number used for evaluating the statistical significance of the difference between observed data and the data that would be expected under a specific hypothesis. The chi-square (χ^2) statistic is defined as a sum of terms of the form

$$\frac{(\text{observed} - \text{expected})^2}{\text{expected}}$$

with one term for each observed value.

Coefficient   Usually, a number used to multiply a variable or power of a variable in an algebraic expression.

Example: In the expression $3x + 4 \sin x$, the numbers 3 and 4 are coefficients.

Completing the square   The technique of adding a constant term to an expression, especially a quadratic expression, to make it a **perfect square.**

Example: To complete the square for the expression $x^2 + 10x$, add 25. The result, $x^2 + 10x + 25$, is the perfect square $(x + 5)^2$.

Congruent   Informally, having the same shape and size. Formally, two polygons are congruent if their corresponding angles have equal measures and their corresponding sides have equal lengths. The symbol $\cong$ means "is congruent to."

Constant term   In any polynomial, a term that does not contain the variable. (If no such term appears in the expression, then the constant term is 0.) See also **polynomial.**

Examples: In the polynomial $x^3 - 2x^2 + 7$, the number 7 is the constant term. In the polynomial $2x^4 + 5x$, the constant term is 0.

Constraint   Informally, a limitation or restriction. In a linear programming problem, any of the conditions limiting the variables.

Cosecant   In a right triangle, the reciprocal of the **sine** ratio. The cosecant of $\angle A$ is abbreviated as csc A. See *A Trigonometric Summary* in the unit *Do Bees Build It Best?*

$$\csc A = \frac{1}{\text{sine } A}$$

Cosine   The ratio of the length of the leg **adjacent** to one nonright angle of a right triangle to the length of the **hypotenuse.** The cosine of $\angle A$ is abbreviated as cos A. See *A Trigonometric Summary* in the unit *Do Bees Build It Best?*

$$\cos A = \frac{\text{adjacent}}{\text{hypotenuse}}$$

Cotangent   In a right triangle, the reciprocal of the **tangent** ratio. The cotangent of $\angle A$ is abbreviated as cot A. See *A Trigonometric Summary* in the unit *Do Bees Build It Best?*

$$\cot A = \frac{1}{\text{tangent } A}$$

Cubic (equation, expression, or function)   An equation, expression, or function involving a polynomial of degree 3. See also **polynomial.**

Degree (of a **polynomial**)   The largest exponent appearing with the variable of a polynomial.

Dependent (equations)   See **system of equations.**

Distributive property   The mathematical principle that states that the equation $a(b + c) = ab + ac$ holds true for all numbers a, b, and c.

Edge   See **polyhedron** or **lateral edge.**

Equivalent equations (or inequalities)   A pair of equations (or inequalities) that have the same set of solutions.

Equivalent expressions   Algebraic expressions that give the same numeric value no matter what values are substituted for the variables.

   Example: $3n + 6$ and $3(n + 2)$ are equivalent expressions.

Expected number   The value that would be expected for a particular data item if the situation perfectly fit the probabilities associated with a given hypothesis.

Exponent   A number written as a superscript to another number or variable (the **base**), to indicate the power to which the base is raised.

Face   See **polyhedron** or **lateral face.**

Factoring   The process of writing a number or an algebraic expression as a product.

   Example: The expression $4x^2 + 12x$ can be factored as the product $4x(x + 3)$.

Feasible region   The region consisting of all points whose coordinates satisfy a given set of **constraints.** A point in this set is a *feasible point.*

Geometric sequence   A sequence of numbers in which each term is a fixed multiple of the previous term.

   Example: The sequence 2, 6, 18, 54, . . . , in which each term is three times the previous term, is a geometric sequence.

Half plane   The set of points in a plane that satisfy a linear inequality.

Height (of a **prism**) A perpendicular segment from one base to the other or the length of this segment.

Hypotenuse The longest side in a right triangle or the length of this side. The hypotenuse is located opposite the right angle.

Hypothesis Informally, a theory about a situation or about how a certain set of data is behaving. Also, a set of assumptions used to analyze or understand a situation.

Hypothesis testing The process of evaluating whether a hypothesis holds true for a given population. Hypothesis testing usually involves statistical analysis of data collected from a sample.

Inconsistent (equations) See **system of equations**.

Independent (equations) See **system of equations**.

Inequality A statement that one expression is less than (or greater than) another.

> *Examples:* $4 > 3$, $4x - 2 < 10$, or $2x - 3 \geq 5x + 1$

An inequality using $<$ or $>$ is a *strict inequality*. An inequality using $\geq$ or $\leq$ is a *nonstrict inequality*.

Inverse trigonometric function Any of six functions used to determine an angle if the value of a trigonometric function is known.

> *Example:* For x between 0 and 1, the inverse sine of x (written $\sin^{-1} x$) is defined to be the angle between 0° and 90° whose sine is x.

Irrational number A number that cannot be expressed as the ratio of two integers.

> *Examples:* $\sqrt{2}$ and π are irrational numbers.

Lateral edge A line segment connecting a vertex of one base of a prism to the corresponding vertex of the other base. See *The World of Prisms* in the unit *Do Bees Build It Best?*

Lateral face A face of a prism other than its bases. See *The World of Prisms* in the unit *Do Bees Build It Best?*

Lateral surface area   The amount of area that the lateral faces of a prism contain. See *The World of Prisms* in the unit *Do Bees Build It Best?*

Law of repeated exponentiation   The mathematical principle that states that the equation

$$(A^B)^C = A^{BC}$$

holds true for all numbers A, B, and C (as long as the expressions are defined).

Leading term (of a polynomial)   The term of a polynomial involving the highest power of the variable.

> *Example:* In the polynomial $3x^2 + 7x^5 + 6$, $7x^5$ is the leading term because 5 is the highest exponent used for x.

Leg   Either of the two shorter sides in a right triangle. The two legs of a right triangle form the right angle of the triangle.

Linear equation   An equation whose graph is a straight line. More generally, an equation stating that two **linear expressions** are equal.

Linear expression   For a single variable x, an expression of the form $ax + b$, where a and b are any two numbers, or any expression equivalent to an expression of this form. For more than one variable, any sum of linear expressions in those variables (or an expression equivalent to such a sum).

> *Examples:* $4x - 5$ is a linear expression in one variable; $3a - 2b + 7$ is a linear expression in two variables.

Linear function   For functions of one variable, a function whose graph is a straight line. More generally, a function defined by a **linear expression.**

> *Example:* The function g defined by the equation $g(t) = 5t + 3$ is a linear function in one variable.

Linear inequality   An inequality in which both sides of the relation are **linear expressions.**

> *Example:* The inequality $2x + 3y < 5y - x + 2$ is a linear inequality.

Linear programming   A problem-solving method that involves maximizing or minimizing a **linear expression,** subject to a set of constraints that are **linear equations** or **inequalities.**

Logarithm The power to which a given **base** must be raised to obtain a given numeric value.

Example: The expression $\log_2 28$ represents the solution to the equation $2^x = 28$. Here, "log" is short for *logarithm,* and the whole expression is read "log, base 2, of 28."

Mathematical model A mathematical description or structure used to represent how a real-life situation works.

Monomial A **polynomial** with just one **term.**

Net A two-dimensional figure that can be folded to create a three-dimensional figure.

Example: The figure on the left is a net for the cube.

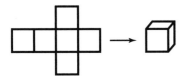

Normal distribution A special case of a bell-shaped distribution with a precise mathematical definition. This distribution can be used as an approximation of real-world situations. See *Normal Distribution and Standard Deviation* in the unit *Is There Really a Difference?*

Null hypothesis A "neutral" assumption of the type that researchers often adopt before collecting data for a given situation. The null hypothesis often states that there are no differences between two populations with regard to a given characteristic.

Order of magnitude An estimate of the size of a number based on the value of the exponent of 10 when the number is expressed in **scientific notation.**

Example: The number 583 is of the second order of magnitude because it is written in scientific notation as $5.83 \cdot 10^2$, using 2 as the exponent for the base 10.

Parabola The type of curve that occurs as the graph of a **quadratic function.** The maximum or minimum point of the graph is the **vertex** (or *turning point*) of the parabola.

Examples: The graphs of the equations $y = x^2$ and $y = -x^2 + 2x + 2$, shown here, are both parabolas. The first is described as "opening upward," and its vertex is at its minimum point, (0, 0). The second is described as "opening downward," and its vertex is at its maximum point, (1, 3).

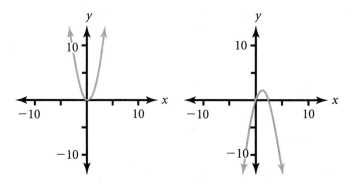

Parallelogram   A quadrilateral in which both pairs of opposite sides are parallel.

Examples: Polygons *ABCD* and *EFGH* are parallelograms.

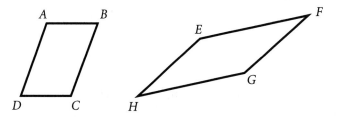

Percentage growth   The proportional rate of increase of a quantity, usually over time, found by dividing the absolute growth in the quantity by the initial value of the quantity. Used in distinction from **absolute growth.**

Perfect square   A number or expression that is the square of a number or expression.

Examples: 49 and $x^2 + 8x + 16$ are perfect squares because $49 = 7^2$ and $x^2 + 8x + 16 = (x + 4)^2$.

Polygon   A closed two-dimensional shape formed by three or more line segments. The line segments that form a polygon are its *sides*. The endpoints of these segments are called *vertices* (or singular **vertex**).

Examples: All of these figures are polygons.

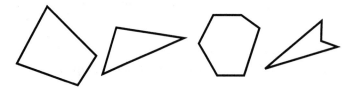

Polyhedron A three-dimensional figure bounded by intersecting planes. The polygonal regions formed by the intersecting planes are the *faces* of the polyhedron, and the sides of these polygons are the *edges* of the polyhedron. The points that are the vertices of the polygons are also vertices of the polyhedron.

Example: This figure is a polyhedron. Polygon *ABFG* is one of its faces, $\overline{CD}$ is one of its edges, and point *E* is one of its vertices.

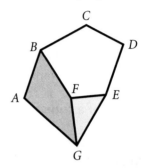

Polynomial (in a given variable) Any expression that is a sum in which each term is a whole-number power of the variable multiplied by a coefficient. A **monomial** is a polynomial with just one term, a **binomial** is a polynomial with exactly two terms, and a **trinomial** is a polynomial with exactly three terms. The **degree** of a polynomial is the highest exponent appearing with the variable.

Examples: $4x^3$ is a monomial of degree 3; $3x^4 - 2x$ is a binomial of degree 4; $-2x^5 + x^2 + 7$ is a trinomial of degree 5. The constant polynomial 7 is a monomial of degree 0.

Population A set (not necessarily of people) referred to in a statistical study and from which a **sample** is drawn.

Prism A polyhedron in which two of the faces are parallel and congruent. For details and related terminology, see *The World of Prisms* in the unit *Do Bees Build It Best?*

Profit line In the graph used for a linear programming problem, a line representing the number pairs that give a particular profit.

Pythagorean theorem The principle for right triangles that states that the sum of the squares of the lengths of the two legs equals the square of the length of the hypotenuse.

Example: In right triangle *ABC* with legs of lengths *a* and *b* and hypotenuse of length *c*, the Pythagorean theorem states that $a^2 + b^2 = c^2$.

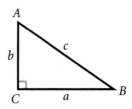

Quadratic (**equation, expression, or function**)   An equation, expression, or function involving a polynomial of degree 2. See also **polynomial.**

Regular polygon   A polygon with sides that all have equal length and angles that all have equal measure.

Right prism   A prism where the direction through which the base has moved is perpendicular to its original position. An ordinary box is an example of a right rectangular prism.

Root (**of a function**)   An input value for a function that creates an output of 0.

Root (**of a number**)   The *square* root of a number is the value that must be squared to give that number; the *cube* root of a number is the value that must be cubed to give that number, and so on for any *n*th root.

   Examples: The square root of 25 is 5 because $5^2 = 25$. The fourth root of 81 is 3 because $3^4 = 81$.

Sample   A selection taken from a **population,** often used to make conjectures about the entire population.

Sampling fluctuation   Variations in data for different **samples** from a given **population** that occur as a natural part of the sampling process.

Scientific notation   A method of writing a number as the product of a number between 1 and 10 and a power of 10.

   Example: In scientific notation, the number 3158 is written as $3.158 \cdot 10^3$.

Secant   In a right triangle, the reciprocal of the **cosine** ratio. The secant of $\angle A$ is abbreviated sec A. See *A Trigonometric Summary* in the unit *Do Bees Build It Best?*

$$\sec A = \frac{1}{\text{cosine } A}$$

Sine   The ratio of the length of the leg opposite one nonright angle of a right triangle to the length of the hypotenuse. The sine of $\angle A$ is abbreviated sin A. See *A Trigonometric Summary* in the unit *Do Bees Build It Best?*

$$\sin A = \frac{\text{opposite}}{\text{hypotenuse}}$$

Standard deviation   A specific measurement of how spread out a data set is, usually represented by the lowercase Greek letter sigma (σ). See *Normal Distribution and Standard Deviation* in the unit *Is There Really a Difference?*

Standard form (of a **quadratic equation**)   A quadratic equation written in the form $y = ax^2 + bx + c$, where a and b are not zero.

Surface area   The amount of area that the surfaces of a three-dimensional figure contain.

System (of **equations**)   A set of two or more equations considered together. If the equations have no common solution, the system is *inconsistent*. Also, if one of the equations can be removed from the system without changing the set of common solutions, that equation is dependent on the others, and the system as a whole is also *dependent*. If no equation is dependent on the rest, the system is *independent*. In the case of a system of two linear equations with two variables, the system is *inconsistent* if the graphs of the two equations are distinct parallel lines, *dependent* if the graphs are the same line, and *independent* if the graphs are lines that intersect in a single point.

Tangent   The ratio of the length of the leg opposite one nonright angle of a right triangle to the length of the leg adjacent to the same angle. The tangent of $\angle A$ is abbreviated as tan A. See *A Trigonometric Summary* in the unit *Do Bees Build It Best?*

$$\tan A = \frac{\text{opposite}}{\text{adjacent}}$$

Term (of a **polynomial**)   Each part of a polynomial separated by addition or subtraction.

Tessellation   A pattern of shapes (often identical) that fit together without overlapping or leaving gaps.

Trapezoid   A quadrilateral in which one pair of opposite sides is parallel and the other pair is not. The parallel sides are the **bases** of the trapezoid.

Examples: Polygons *KLMN* and *PQRS* are trapezoids.

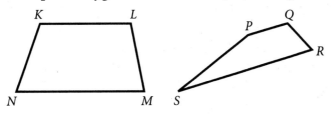

Trigonometry For a right triangle, the study of the relationships among the acute angles and the lengths of the sides. For details, see *A Trigonometric Summary* in the unit *Do Bees Build It Best?*

Trinomial A **polynomial** with exactly three **terms.**

Unit fraction A fraction in which the numerator is 1 and the denominator is a positive integer.

Vertex (for a parabola) The minimum or maximum point of the graph.

Vertex (for a polygon or polyhedron) An endpoint of a side or edge. See also **parabola, polygon,** and **polyhedron.**

Vertex form (of a quadratic equation) A quadratic equation written in the form $y = a(x - h)^2 + k$, where a is not zero. The point (h, k) is the vertex of the parabola.

Volume A measure of the amount of space contained in a three-dimensional figure.

x-intercept A point where a graph crosses the x-axis. Sometimes, the x-coordinate of that point.

Zero product rule If the product of two numbers is zero, then at least one of the numbers in the product must be zero.

INDEX OF MATHEMATICAL IDEAS

The phrase "concept development" indicates that the associated term is previewed or developed on the page, but may not be named.

B

base (of a triangle, parallelogram, or trapezoid), 412

base (of an exponent or logarithm), 389, 412

bases (of a prism)
  defined, 64, 412
  concept development, 68, 69

bell curve. *See* normal distribution

bias
  defined, 412
  concept development, 193–194, 196, 197, 234–235, 238

binomials, 329–330, 412

C

calculators
  random number generators, 233
  "$y =$" form of equations for, 150

chi-square (χ^2) distribution, concept development, 230, 231, 232, 233

chi-square (χ^2) statistic
  computation of, 224
  defined, 412
  degrees of freedom. *See* degrees of freedom
  and factors of ten, 226
  linear interpolation and, 267–268
  meaning of, 224
  notation for and pronunciation of, 213
  probability tables, 234–235, 267–268, 275–276
  sample size and, 224–225, 266
  theoretical models and, 248
  concept development, 220, 223, 227, 228, 230, 231, 232, 233, 234–235, 236, 247–248, 249, 250, 251–252, 253, 254, 255–256
  See also statistical reasoning

circles
  as base of cylinder, 70
  radius, 50, 106
  concept development, 50

coefficients
  constant term as, 328
  defined, 412
  inequalities and, 122
  leading, 330
  of polynomials, 329

completing the square
  defined, 413
  process of, 301
  concept development, 300, 310, 311, 312, 313, 314, 315
  See also quadratic expressions

concavity
  of normal curve, 216
  of parabolas, 291, 329

conditional probability, concept development, 221–222, 265

congruence
  defined, 413
  concept development, 8–9, 12, 38–40, 50

conjecture, concept development, 49, 157–158

constant term
  as coefficient, 328
  defined, 413
  degree of, 330
  of polynomial, 329
  as polynomial, 330

constraints
  defined, 413
  feasible regions. *See* feasible regions
  graphing, 126–127, 132
  as inequalities, 124
  concept development, 116–117, 118, 128, 131, 137, 139–140, 141, 145, 146, 147, 148, 149, 161–162, 165, 168–169, 170

conventions, 366

coordinate geometry
  concept development, 126–127, 132, 133–134, 139–140, 141, 153
  concept development: reflections, 14, 389
  concept development: rotation, 8–9, 14

coordinate system
  plotting points on, 126–127, 132, 137
  scales of axes, 360
  concept development, 133–135, 136,
    139–140, 141, 153, 285, 286, 288,
    289, 290, 291, 310, 361, 377–378, 389
  See also graphs and graphing
cosecant, 23, 413
cosine, 23, 413
cosines, law of, 102
cotangent, 23, 413
cubic functions, 348, 413
cubic polynomials, 330
cylinders, concept development, 69, 111

D

data
  spread, 214
  table for set of, 217
  concept development: inference,
    193–194, 206, 210, 211, 212
  concept development: one-variable,
    214–217, 218
data representations. *See* graphs and
  graphing
decagons, 69
deduction, 358
deductive reasoning, concept development,
  15, 18–19, 36, 129–130, 132, 139–140,
  143–144, 153, 178, 302, 358–359, 368,
  376, 379, 381–382, 398–399, 402
degree of a polynomial, 330, 414
degrees of freedom
  calculation of, 272–273, 274
  of m-by-n table, 272–274
  table of probability for chi-square
    statistic, 275–276
  of 2-by-2 table, 272, 275
  concept development, 269–270
dependent systems of equations
  defined, 421
  concept development, 159, 163, 172
depreciation, 397

design of statistical experiments
  quality of, 197, 199–200
  concept development, 193–194, 196,
    201–202, 203, 204, 205, 206, 234–235,
    251–252, 254, 264
difference of squares, 348
distributions
  normal. *See* normal distribution
  concept development: chi-square (χ^2),
    230, 231, 232, 233
  concept development: probability
    along, 80–82, 83–86, 212, 230, 231
distributive property
  algebraic expressions, multiplication
    of, 298
  area models, multiplication with, 297,
    299
  defined, 296, 414
  multidigit multiplication and, 296–297
  concept development, 150, 154, 159,
    294, 295, 300, 310, 321, 322, 323, 335
division
  inequalities and, 119
  square roots and, 49
divisors, counting, 338
dodecagons, 69
double bar graphs, concept development,
  193–194, 201–202

E

edges
  defined, 419
  lateral, 65, 415
elevation, angle of, 29
equations
  equivalent. *See* equivalent equations
  finding x-intercepts as equivalent to
    solving, 323
  linear. *See* linear equations
  quadratic. *See* quadratic equations
  true, 119
  "$y =$" form of, 150

partial sums, 297

pentagons, area and perimeter, relationship of, 48

percentage growth
   defined, 418
   concept development, 360, 397
   See also rate of change

perfect square
   completing the square for, 301
   defined, 418
   twin primes and, 302

perimeter
   area and, relation of, 45, 46, 47, 50, 70, 340
   semiperimeter, 105, 340

plotting points, 126–127, 131, 132, 137, 142

polygons
   area of, 11, 98
   defined, 418
   n-gons, 50
   naming of, 16
   prisms and, 64
   regular. *See* regular polygons
   tessellations with nonregular, 112
   concept development, 21, 132, 136, 137, 139–140

polyhedrons
   defined, 419
   nets for, 57, 58
   volume and surface area of, 59–60
   concept development, 6, 64–65, 66–67, 68

polynomials
   binomials, 329–330, 412
   coefficients of, 329
   constant term alone, 330
   constant terms of, 329
   cubic, 330
   defined, 329, 419
   degree of, 330, 414
   missing terms of, 330
   monomials, 329, 417
   trinomials, 328–330, 422
   concept development, 347

population
   defined, 419
   concept development, 193–194, 195, 196, 197, 199–200, 201–202, 247–248, 249, 251–252, 265, 269–270, 272–273
   See also samples

prime numbers
   defined, 302
   and divisors, finding, 338
   twin primes, properties of, 302

prisms
   bases of. *See* bases (of a prism)
   defined, 419
   height of. *See* height of a prism
   hexagonal, 64
   lateral edges of, 65, 415
   lateral faces of, 65, 415
   lateral surface area of. *See* lateral surface area
   oblique, 65
   rectangular, 64
   right. *See* right prisms
   summary of, 64–65
   triangular, 64
   volume and surface area of, 59–60
   volume of, 62, 68, 69, 111
   concept development, 6, 57, 58, 66–67, 68, 69

probability
   conditional, 221–222, 265
   estimating with frequency bar graph, 232
   expected number. *See* expected number
   concept development: along distributions, 80–82, 83–86, 212, 230, 231
   concept development: independent events, 83–86
   See also probability tables

probability tables, 234–235, 267
   degrees of freedom, 275–276

Problems of the Week (POWs), standard write-up for, 26–27

concept development, 139–140, 141,
   392–393
ratio
   Pythagorean theorem proof and, 96–97
   trigonometric. *See* trigonometry
   concept development, 21, 47, 71, 72, 195
rational numbers, concept development, 380
rays, notation for, 17
rectangles, 16
rectangular prisms, 64
reflections, concept development, 14, 389
regular polygons
   area and perimeter, relationship of, 48,
      50, 340
   defined, 420
   tessellating, 75
   concept development, 48, 69, 111
Richter scale, 406–407
right prisms
   defined, 65, 420
   concept development, 66–67, 69
right triangles
   concept development, 20, 24–25, 31,
      32, 33, 34, 35, 36, 37, 41, 42, 48, 50,
      52, 61, 66–67, 68, 94, 95, 96–97, 102,
      103, 308, 309
   See also Pythagorean theorem;
      triangles; trigonometry
roots (of a function)
   defined, 420
   finding, 329
   See also *x*-intercepts
roots (of numbers)
   defined, 420
   exponents and, 376, 408
   extraneous solutions, 342–343
   See also exponents; square roots
rotation, concept development, 8–9, 11, 14
rounding numbers
   decimals, 35
   maximum and, 107

S

sample size, chi-square statistic and,
   224–225, 266

samples
   choosing, 193–194
   defined, 420
   concept development, 195, 196, 197,
      201–202, 234–235
sampling fluctuation
   defined, 420
   evaluation of, 213
   concept development, 193–194, 195,
      196, 199–200, 205, 210, 212, 247–248,
      251–252
scale factor
   defined, 73
   concept development, 71, 72
scientific notation
   and big numbers, 392–393
   defined, 420
   use of, 391
   concept development, 409
secant, 23, 420
semiperimeter, 105, 340
sequences, 400–401
similarity
   area and, 98
   defined, 22
   Pythagorean theorem and, 96–97
   concept development, 20, 24–25, 80–82,
      83–86
simulations
   coin toss, 230
   random number generators, 233
   concept development, 204, 205, 206,
      207–208, 229, 231, 232
sine, 23, 420
sines, law of, 103
slope. *See* rate of change
spread, 214
square roots
   extraneous solutions, 342–343
   functions of, operations on, 49
   as length, 35
squares, area and perimeter of, 46, 47
squaring function, interpolation and, 268
standard deviation
   calculation of, 216–217

theoretical models, chi-square statistic and, 248

time, finding, 310

transformations
    scale factor. *See* scale factor
    concept development: reflections, 14, 389
    concept development: translations, 286, 288, 289

translations, concept development, 286, 288, 289

trapezoids
    altitude of, 411
    area of, 20
    defined, 421
    naming, 16

triangle inequality, concept development, 34

triangles
    altitudes of. *See* altitudes
    equilateral. *See* equilateral triangles
    isosceles. *See* isosceles triangles
    right triangle trigonometry. *See* trigonometry
    concept development: angle sum property, 21, 61
    concept development: right, 20, 24–25, 31, 32, 33, 34, 35, 36, 37, 41, 42, 48, 50, 52, 61, 66–67, 68, 94, 95, 96–97, 102, 103, 308, 309
    See also Pythagorean theorem; triangle inequality

triangular prisms, 64

trigonometry
    and altitudes, height of, 28
    defined, 422
    inverse functions. *See* inverse trigonometric function
    as ratios, 22–23
    summary of, 22–23
    concept development, 20, 21, 24–25, 29, 47, 48, 50, 52, 61, 102, 103

trinomials, 329, 330, 422

true equations or inequalities, 119

twin primes, 302

U

unit fractions, 380

units of measure
    for length, 62
    for surface area, 62
    for volume, 62
    concept development, 8–9, 10, 13, 32, 59, 61, 62, 90, 91, 284
    See also measurement

V

vertex form
    and coefficients greater than 1, 346
    conversion to standard from, 300, 303–304
    defined, 289, 422
    focal length and, 304
    graphing of, 289, 290
    parameters of, 289
    and x-intercepts, finding, 310
    and x-intercepts, number of, 291
    concept development, 293, 301, 311, 312, 313, 314, 315, 317, 346, 351

vertex (of a parabola)
    defined, 285, 422
    and focal length, 304
    line of symmetry through, 333
    as minimum or maximum, 329
    position of, and number of x-intercepts, 291
    concept development, 286, 288, 289, 290, 300, 310, 311, 312, 314, 315, 317, 324, 325, 336, 346, 347, 350, 351

vertex (vertices) of a polygon or polyhedron
    defined, 422
    naming of polygons by, 16
    concept development, 12, 21, 66–67, 96, 102

vertical-multiplication form, 299

volume
    and area, relation of, 58, 59–60, 63, 69
    of cylinders, 70, 111
    defined, 422
    estimation of, 88–89

maximum and, 108–109, 110, 111
of prisms, 62, 68, 69, 111
and scale factors, 73
concept development, 4, 6, 61, 63, 71, 72

INDEX OF ACTIVITY TITLES

PHOTOGRAPHIC CREDITS

Front Cover Photos

(From top left, clockwise) Kelsey Agardy, Antonio Cruz, Nicholas Birago, Justin Choate, Morris Norrise, Joshua Tovar, Elana Cohen, Sam Regan, Alyssa Spediacci

Front Cover and Unit Opener Photography

Berkeley High School and Lincoln High School: Stephen Loewinsohn

Bees

3 Lincoln High School, Stephen Loewinsohn; **5** The Image Bank; **7** Lincoln High School, Stephen Loewinsohn; **26** Lincoln High School, Stephen Loewinsohn; **30** Lincoln High School, Stephen Loewinsohn; **34** Shutterstock; iStockphoto; iStockphoto; **44** Pleasant Valley High School, Mike Christensen; **47** Shutterstock; **53** Foothill High School, Cheryl Dozier; **73** Michael Newman, PhotoEdit; **74** Lincoln High School, Stephen Loewinsohn; **80** PhotoDisc; **81** PhotoDisc; **87** CMCD

Cookies

115 Berkeley High School, Cheryl Fenton; **116** KCP; **118** Hillary Turner; **124** KCP; **125** Colton High School, Sharon Taylor; **138** Berkeley High School, Cheryl Fenton; **148** Comstock ©1995; **151** iStockphoto; **152** Berkeley High School, Cheryl Fenton; **155** Getty Images; **160** Capuchino High School, Chica Lynch; **162** KCP; **167** Berkeley High School, Cheryl Fenton; **168** iStockphoto; **169** KCP; **170** Berkeley High School, Stephen Loewinsohn; **171** Berkeley High School, Cheryl Fenton; **174** Berkeley High School, Cheryl Fenton; **175** Berkeley High School, Stephen Loewinsohn; **184** Comstock ©1996

Is There Really a Difference?

191 Santa Maria High School, Mike Bryant; **197** Tony Freeman, PhotoEdit; **198** Lincoln High School, Stephen Loewinsohn; **202** Bob Daemmrich, PhotoEdit; **212** Lincoln High School, Stephen Loewinsohn; **213** Lincoln High School, Stephen Loewinsohn; **218** Comstock ©1993; **227** Comstock ©1995; **233** PhotoDisc; **237** Lincoln High School, Stephen Loewinsohn; **240** Lincoln High School, Stephen Loewinsohn; **249** iStockphoto; **250** Corbis; **257** Foothill High School, Cheryl Dozier; **259** Berkeley High School, Stephen Loewinsohn; **260** Lincoln High School, Stephen Loewinsohn; **265** iStockphoto; **266** Michael Newman, PhotoEdit; **270** Bonnie Kamin, PhotoEdit

Fireworks

279 Berkeley High School, Stephen Loewinsohn; **280** The Image Bank; **282** PhotoDisc; The Image Bank; **283** Stock Boston; **287** Animals, Animals; **293** Berkeley High School, Stephen Loewinsohn; **307** Mendocino High School, Don Cruser and Lynne Alper; **308** PhotoDisc; **312** Science Photo Library; **315** Getty Images; **316** Berkeley High School, Stephen Loewinsohn; **317** The Image Bank; **318** Getty Images; **319** Shutterstock; **320** Berkeley High School, Stephen Loewinsohn; **324** PhotoDisc; **326** Berkeley High School, Stephen Loewinsohn; **327** Berkeley High School, Stephen Loewinsohn; **332** Laura Murray; **340** The Image Bank; **346** PhotoDisc

All About Alice

355 Lincoln High School, Stephen Loewinsohn; **358** Archive Photos; **361** Douglas Engle, Corbis; **362** San Lorenzo Valley High School, Dennis Cavaillé; **375** Lincoln High School, Stephen Loewinsohn; **386** Lincoln High School, Stephen Loewinsohn; **392** iStockphoto; **394** Berkeley High School, Stephen Loewinsohn; **406** Jim Sugar, Corbis; **408** Rose Hartman, Corbis; **409** Getty Images; PhotoDisc